Small States in Multilateral Economic Negotiations

This book addresses the puzzle, *Can David take on Goliath in multilateral economic negotiations, and if so, then under what conditions?* The question of how the weak bargain with the strong in international politics is exciting theoretically and empirically. In a world of ever-increasing interdependence, and also a time of economic crisis, it acquires even greater significance.

With the help of issue-specific case studies, the volume offers new insights into the vulnerabilities that small states face in multilateral economic negotiations, and also mechanisms whereby these weaknesses might be overcome and even used as an advantage. The attention that this volume pays to questions of smallness and negotiation allows it to address a long-standing problem of international politics. The case studies, which cover monetary, financial, trade, and climate change negotiations, ensure a unique and valuable topicality to the volume.

This book was published as a special issue of *The Round Table*.

Amrita Narlikar is Reader in International Political Economy and Director of the Centre for Rising Powers at the University of Cambridge. Her latest books include *The Oxford Handbook on the WTO* (co-edited), Oxford University Press, 2012, and *New Powers: How to become one and how to manage them,* Columbia University Press, 2010.

Small States in Multilateral Economic Negotiations

Edited by
Amrita Narlikar

Routledge
Taylor & Francis Group

LONDON AND NEW YORK

Contents

Citation Information

The chapters in this book were originally published in *The Round Table*, volume 100, issue 413 (April 2011). When citing this material, please use the original page numbering for each article, as follows:

Introduction: Small States in Multilateral Economic Negotiations

This journal has a long-standing reputation for publishing pioneering, state-of-the-art research on small states. So it seemed only appropriate that in the series of conferences and workshops that have been organised to commemorate *The Round Table*'s centenary, the first was on Small States in International Economic Negotiations. The workshop was held in November 2009, sponsored by the journal and hosted by the University of Cambridge, and this special issue is a product of the debate that took place there. Its aim is to take stock of research on small states thus far, and focus and develop it further. The central puzzle driving this special issue is how some of the smallest players negotiate in the international political economy, or how the weak bargain with the strong in multilateral economic negotiations.

For the small and vulnerable attempting to influence and secure agreements on favourable (or at least less unfavourable) terms, an analysis of their negotiation strategies is crucial to understanding what works and what does not in their dealings with much larger and stronger counterparts. But the question of their negotiation behaviour is just as important from a theoretical perspective: for scholars of international bargaining and negotiation, small states present a hard test case of whether smart negotiating strategies can make a difference. In this special issue, case studies are drawn on from a diverse set of regimes, ranging over international monetary and financial regimes, trade and climate change, to address collectively the question: Can David successfully take on Goliath in international politics, and if so, then under what conditions?

The four case studies in this collection are preceded by a conceptual essay by Paul Sutton, which addresses the contested issue of the definition of small states. Sutton's paper demonstrates how matters of categorisation and nomenclature are fundamentally political, subject to negotiation, and with potentially profound distributive consequences. Key actors in this debate are not just the small states themselves, but several international organisations. These include the Commonwealth Secretariat and the World Bank (which Sutton identifies as the 'champions' of small states), but also others such as the World Trade Organisation (WTO) and the United Nations Conference on Trade and Development (UNCTAD). Sutton points to the difficulties of defining 'small', but concludes by honing in on the two criteria employed by the Commonwealth Secretariat and the World Bank: population size (1.5 million or below in most cases) and vulnerability. He writes, '...vulnerability should be seen as the core characteristic of small states in the contemporary international political economy. It sets them apart from most other states and establishes an agenda in many ways unique to their needs.'

The vulnerabilities that small states face in their international negotiations receive considerable attention from several of the contributors to this special issue. While the notion of vulnerability is in some ways intrinsically associated with smallness, the post-Cold War system provides at least three major and additional reasons for concern. First, the end of the Cold War and the decline of competing economic ideologies have meant that no state can risk opting out of the increasingly integrated economic system. Even large developing countries today have learnt to embrace this as they recognise that the costs of opting out outweigh the costs of joining in. For smaller developing countries, the costs of staying out are higher still, given that their BATNA, i.e. best alternative to negotiated agreement, is even more limited, and their dependence on international markets means that they have almost no go-it-alone power (Gruber, 2001). Second, to the extent that the rivalry between the two superpowers allowed 'big influence' to 'small allies' (Keohane, 1971), the end of the Cold War has deprived small states of an important source of bargaining leverage. Third, as the larger developing countries have risen to greater power, evidenced for instance in the inclusion of Brazil and India in the 'New Quad' or 'Core Group' in the WTO, they have begun to attract considerably greater scholarship than other developing countries. Their rise entails other risks for smaller states besides academic neglect. Traditionally, some of the larger developing countries had led coalitions involving smaller developing countries, and had assisted in ensuring that their demands were placed on the table. Today, there is a danger that as rising powers make their way into the club of Great Powers, they will no longer regard it as worthwhile to fight the cause of the smaller countries. Indeed, disgruntled mutterings by smaller countries against Brazil and India are audible in the corridors of the WTO (Ismail, 2009; Odell, 2010). Small states now risk finding themselves friendless and alone at a time when they are much more vulnerable as a result of increasing integration in the world economy. The sources of their vulnerabilities need to be clearly identified, and also strategies that they could effectively employ to improve their bargaining position.

Three of the four case studies in this special issue highlight the vulnerabilities that small states face in the current context. Two papers focus particularly on the impact that the financial crisis has had on further reducing the negotiating space for small states. André Broome examines the negotiations of small states with the International Monetary Fund (IMF) in times of economic distress. He argues that although smallness can provide states with some important opportunities (e.g. the development of specialist sectors such as banking), the risks inherent in financial integration also greatly increase their vulnerability to external shocks. Using the case of Iceland's negotiations with the IMF in the aftermath of the recent financial crisis, he demonstrates that small states encountering the consequences of 'disaster capitalism' are likely to face some high costs. For example, the IMF's involvement in the rescue is likely to come at the cost of economic sovereignty and associated decline in electoral support at home. Further, attempts to improve BATNA by conducting parallel negotiations with other financiers may actually end up increasing the small states' dependence on the IMF. This is because other states usually make bilateral credit conditional on the recipient state's ability to adhere to the goals of the IMF programme. Effectively, some states may be able to utilise their smallness to their advantage when times are good, but face some very difficult challenges when the financial going gets tough.

The second paper that examines the consequences of the financial crisis on small states is by Mark Hampton and John Christensen, which focuses on small island economies that have specialised in offshore finance. Interestingly, the authors point to how small states had used successful framing tactics in the past to resist initiatives against tax evasion in the early 2000s. For example, in countering the Organisation for Economic Cooperation and Development (OECD) Harmful Tax Competition initiative, small states successfully used the language of 'fiscal colonialism' by the OECD and further framed their fight as one of the small and powerless against the 'cartel' of OECD countries. But in the aftermath of the recent financial crisis, Hampton and Christensen argue that although 'it is unlikely that the initiatives set in motion in 2009 will cause the demise of all tax havens ... There is little doubt that the combination of measures by the OECD and EU will restrict the activities of existing actors and radically reduce their ability to resist requests for international cooperation in tackling tax evasion ...' Evolving context and path dependence have a big impact on the ability of small states to negotiate: the international milieu is less tolerant of tax havens in hard times, and '... the past actions and policy choices of the small island hosts themselves have contributed significantly to the serious predicament that they now find themselves in, and consequently the extremely limited economic development possibilities that remain open to them'.

The third case study is by Brendan Vickers on small states in trade negotiations. He focuses particularly on small states involved in negotiations between the EU and the South African Development Community (SADC) as part of the process of concluding the Economic Partnership Agreements (EPAs) between the EU and the ACP (African, Caribbean and Pacific) group of countries. Vickers argues, 'Unlike most other ACP EPA negotiations, SADC's small states have been caught between a rock (EU) and a hard place (South Africa)'. He attributes the difficulties that small states have encountered in the negotiation to several reasons. Major players in the negotiation from the SADC side have been guided by 'domestic interests over regional coherence and collective representation' (with the latter being essential preconditions if the interests of the small and weak are to be safeguarded). South Africa, in particular, 'appears to have shown insufficient regard for its smaller neighbours' peculiar challenges'. Add to this the complications and inadequacies of the negotiation process itself, which Vickers traces carefully, and it is not surprising that small states have had a raw deal.

The dangers that small states face in multilateral economic negotiations are thus plentiful, but as all the papers in this issue recognise, there is also some scope for agency. In line with this, one of Vickers's conclusions is that 'SADC's trade diplomats were not hapless victims of the EU's mercantilist machinations. Instead, their own disarray was partly responsible for the region's undoing. In other words, judicious agency still matters, particularly for small states.' Hampton and Christensen, while highlighting the problems that small island economies face under a reformed system of tightened international financial regulation, also illustrate how they had earlier capitalised on the previous regime of weaker regulation and successfully used their smallness to fight against the last wave to curb tax evasion. These findings add to the small but rich scholarship that recognises that smallness, under certain conditions, need not be a disadvantage and may even be a source of strength. At a minimal level, a small state can benefit from what Amstrup (1976) calls the 'Small State Paradox': if the existence of a small state is not contested by any of the great powers, then its problem of survival also becomes less acute. In the field of economic negotiation, this advantage would translate into small states being unlikely targets of major dispute actions by large states; it is also unlikely that the major players would subject them to the pressures that they usually impose on larger developing countries to make concessions in trade negotiations. A classic illustration of Mancur Olson's argument about 'the "exploitation" of the *great* by the *small*' (Olson, 1965, p. 3) was seen in the operation of the Principal Supplier Principle in the General Agreement on Tariffs and Trade (GATT), when smaller players enjoyed the benefits of tariff reductions negotiated by the major players without having to make reciprocal concessions. As John Odell has argued in a recent article, small states have greater opportunity to serve as chairs of negotiating bodies as 'this mediator function requires a widespread confidence that the information shared with the chair will not be used against us. A delegate from the United States or EU might be highly respected, but the member has a large range of interests and lobbies capable of bringing pressure on the home government' (Odell, 2010, p. 562). Richard Benwell's paper in this issue—the fourth case study—builds on this literature on the advantages of smallness, and takes it further, through an analysis of how small states have emerged as influential norm entrepreneurs in climate change.

Benwell starts off by noting, '...while small states can keep above the water, it is still assumed that they cannot turn the tide of international events'. In the issue area of climate change mitigation, small states present a particularly hard test case for agency: the urgency of the risks that they face (in comparison with larger states) and the necessity of collective action to mitigate climate change render them even more vulnerable than in other issue areas. Benwell recognises that the successes of the small states in securing their value-claiming objectives of financing and technology have been limited, but he persuasively argues that to focus only on such objectives would be too narrow; rather, the success of small states in the climate change regime needs to be seen in terms of overall mitigation action. He identifies their principal source of strength as the 'power of exhortation' that relies on an appeal to scientific as well as moral principles, and their ability to frame their demands as 'the principal victim of a common resource problem not of their own making, small states' power lies in their powerlessness'. Beyond creating awareness of the necessity of mitigation, Benwell further points to concrete gains that small states have managed to achieve through effective negotiation strategies. These include securing recognition for Small Island Developing States as a distinct category in the United Nations, a special seat on the COP Bureau, and also positions of responsibility as chairmen and vice-chairmen in negotiation forums. Benwell's paper provides us with an excellent illustration of how limitations of hard power notwithstanding, small states can exercise considerable influence in the international political economy.

The papers in this issue together offer new insights into the vulnerabilities of small states, particularly in the aftermath of the financial crisis, but some of the papers also offer grounds for cautious optimism. In some instances, small states have been able to overcome their weaknesses. In others, they have gone a step further and converted their smallness into a source of strength. Even in accounts where small states have ended up with unfavourable results, the authors identify sources of strength that remained unexploited in the negotiation that may still offer some bargaining leeway in the future. This is good news for the small states themselves, but also for analysts of negotiation behaviour who seek to explore the conditions under which effective bargaining (even by some of the weakest and most vulnerable players in the political economy) can alter international outcomes.

Acknowledgements

Thanks are extended to all the speakers, discussants and participants at the workshop for their valuable comments, and particularly Terry Barringer, Gordon Baker, Godfrey Baldacchino, Lorand Bartels, Richard Bourne, Andrew Gamble, Dan Kim, Donna Lee, Alex May, James Mayall, Ian Ralby, Cyrus Rustomjee, Ronald Sanders, Nicola Smith and William Vleck.

Amrita Narlikar

References

Amstrup, N. (1976) The perennial problems of small states: a survey of research efforts, *Cooperation and Conflict*, 11(2), pp. 163–182.

Gruber, L. (2001) Power politics and the free trade bandwagon, *Comparative Political Studies*, 34(7), pp. 703–741.

Ismail, F. (2009) Reflections on the WTO July 2008 collapse: lessons for developing country coalitions, in A. Narlikar and B. Vickers (Eds), *Leadership and Change in the Multilateral Trading System* (Leiden: Martinus Nijhoff).

Keohane, R. O. (1971) The big influence of small allies, *Foreign Policy*, 2, pp. 161–182.

Odell, J. (2010) Negotiating from weakness in international trade negotiations, *Journal of World Trade*, 44(3), pp. 545–566.

Olson, M. (1965) *The Logic of Collective Action: Public Goods and the Theory of Groups* (Cambridge, MA: Harvard University Press).

The Concept of Small States in the International Political Economy

PAUL SUTTON

Caribbean Studies Centre, London Metropolitan University, London, UK

ABSTRACT   *This article examines the literature on small states from the related disciplines of international economics and international politics. By accident and design there is no generally agreed definition and characterisation of small states, although those advanced by the Commonwealth Secretariat and World Bank are most satisfactory. The role of the Commonwealth as a champion of small states is examined. Particular attention is paid to the concept of vulnerability and the challenges and opportunities to small states in a globalised world.*

Introduction

The problem of defining and conceptualising small states is illustrated with a personal anecdote. In the spring of 1997 I was appointed by the Commonwealth Secretariat as lead consultant for a new study on small states to update the Commonwealth report *Vulnerability: Small States in the Global Society* (Commonwealth Secretariat, 1985) published 12 years earlier. From the outset I was faced with a problem. The 1985 report defined a small state as an independent country with a population of around one million or less. On that basis 29 Commonwealth states were then defined as small (including both Lesotho and Trinidad and Tobago, which had populations marginally in excess of one million) and were identified as the primary (but not exclusive) subject of the study. If the same criteria were strictly applied in 1997 the around one million cut-off point could reduce the number of states to 26 (excluding Botswana, the Gambia and Mauritius) or even 24 (now excluding Lesotho and Trinidad and Tobago as well). It was clear to me that this would not be politically acceptable so I simply proposed lifting the threshold to one-and-a-half million, arguing in the new report that since 1985 'world population has increased and relative upward adjustment of population figures is necessary to take account of this fact' (Commonwealth Secretariat, 1997, paragraph 2.3). This had the effect of now including these three states plus the 'exceptions' in 1985 of Lesotho and Trinidad and Tobago. The Advisory Group to the report accepted this recommendation without dissent. Equally I retained, again on grounds of political unacceptability were they to be excluded, the arguments in the 1985 report on including states within the immediate geographic region with a population of well above one million on the grounds made then that 'they share many characteristics and also maintain integral links with all small states in their respective regions' (Commonwealth Secretariat, 1985, paragraph 1.5). The result in 1997 was to add in two extra states (Jamaica and Papua New Guinea) to arrive at the same listing as in 1985, with the addition of Namibia, which was now independent. Again the Advisory Group accepted the recommendation without dissent. Indeed, the only dissent on which state was to be included or excluded in definitional terms came from Singapore, who in the preliminary discussions leading to the 1997 study wished at first to be included, but was prevailed upon once the study commenced not to be considered.

The point is made to show how the concept of 'small state' is imprecise and subject to judgements: expedient, informed, or otherwise. This is routinely acknowledged in the literature on small states, in which there is still the absence of a precise definition (Maass, 2009, p. 65) and in some quarters still a search for one (Crowards, 2002). In this article, the focus is on small states in international political economy. It examines some of the arguments on small states encountered in the literature on international economics (especially development economics) and international politics, both of which have contributed to the study of international political economy; and the current debate and understanding of the concept of 'small state' in the contemporary globalised world by officials and policy-makers from states and international agencies, which have influenced the current usage of the concept.

Some Observations on Small States in International Economics

The starting point, and the most cited work in the contemporary study of small states in international economics, is the volume reporting the proceedings of the International Economics Association in 1957. This focused on the 'Economic Consequences of the Size of Nations' and included a comparative study by Kuznets that identified small states as those with a population below 10 million (Kuznets, 1960). It was followed by Demas (1965), who identified a small country as one with a population of five million or less and a useable land area of 10,000–20,000 square miles, and Jalan (1982), who identified it as a population below five million, arable land area below 25,000 square kilometres, and a GNP below US$2bn. He also proposed a subclassification of micro-states with a population of 400,000 or less, arable area of 2,500 square kilometres or less, and GNP below US$500m. These figures suggest both a downward movement in the definition of what constitutes a small state and an attempt to be more precise about definition.

The downward movement in size was a direct reflection of the flood of new 'small states' as a consequence of decolonisation, with 11 with a population of around one million or less admitted to UN membership in the 1960s, 17 in the 1970s, and eight more in the 1980s. The consequence was a rise in concern with small states in the UN and the development of a new interest in them, particularly in their economic development. It was represented in new studies that advocated the concept of the 'micro-state', defined as a state with a population of less than one million (Hein, 1985). Increasingly, in practice this became the 'default definition' by international agencies such as the UN, the World Bank and the Commonwealth Secretariat, although the term 'micro-state' was dropped in favour of 'small state'.

The other exercise was an attempt to be more precise about the definition of a 'small state'. An early attempt by Charles Taylor (1969) used the statistical technique of cluster analysis to identify a group of 74 micro-states (including states and non-independent territories) in which the upper limit was a population of less than 2,928,000, a land area of less than 142,888 square kilometres and a GNP less than US$1,583m. Other studies (Downes, 1988; Downes and Mamingi, 2001) followed that showed there was a close correlation between the three variables used, allowing the choice of population as proxy for the other two, but no final definition of population size emerged. As such, economists have been free to choose their own. For example, papers presented at the 1998 IESG mini-conference 'Small States in the International Economy' showed a variation in population size for 'small states' from 10 million downwards to one million. The same variation in population size was repeated in the Cambridge Workshop on Small States in November 2009, at which this paper was originally presented. In such circumstances it is clearly difficult to determine what economic characteristics small states hold in common yet alone develop a policy to encompass their interests. It is also somewhat uncharacteristic of economics as a discipline, which usually has greater conceptual clarity.

The sort of problems this gives rise to is illustrated by the issue of 'islandness'. If a small state (or micro-state) is defined as possessing a population of one million or less then the majority are islands. In 1972, the United Nations Conference on Trade and Development (UNCTAD) set in motion studies that first gave birth to a special programme on 'land-locked and island developing countries' and then, for island states, morphed into the category of small island developing states (SIDS) between 1992 and 1994 (Hein, 2004). This new concept at least had the benefit of being more specific to small states because 'large' island states had been included in the earlier studies, but it did not resolve the issue identified in a 1983 UNCTAD Working Paper of whether 'smallness rather than insularity is the dominant factor in determining the specific problems of island developing countries' (UNCTAD, 1983, paragraph 7). Rather, terms such as 'remoteness' and 'constraints in transport and communication' seeped into the literature as typical characteristics shaping the economic develop-ment of small states when they should properly have been identified as those of small *island* states. By contrast, the work on small states *qua* small states undertaken by the Commonwealth Secretariat avoided this confusion, listing the economic character-istics of small states as: limited domestic opportunities leading to openness and susceptibility to adverse developments elsewhere; a narrow resource base leading to specialisation in a few products with associated export concentration and dependence on a few markets; shortage of certain skills and high per capita costs in providing government services; and greater vulnerability to natural disasters and greater reliance on overseas aid and various preferential agreements (Common-wealth Secretariat, 1996, p. 1). At the same time, however, and in the same publication, no explicit definition of 'small state' was offered and the country tables to which these remarks applied included, as small states, those with populations of five million or less. So, at the end of several decades of economic research on small states it is fair to comment that some confusion still applies to the definition and the characteristics of such states, even though some studies have attempted to isolate islandness and size as explanatory factors (Armstrong and Reid, 2006, pp. 141–145).

Some Observations on Small States in International Politics

The problem of definition has also permeated the study of small states in international politics. At first this seems odd, because for much of recorded history small states were the majority in international politics throughout the world and large states (almost invariably empires) the few. Yet there is a relatively simple explanation for this fact. It lies in the understanding of what is presumed to be the central dynamic of international politics—the struggle for power—and the confusion of weak states with small states.

The term 'small states', as Neumann and Gstohl (2004, p. 3) pointed out in their review of 'small states in international relations', is relatively new, whereas the concept of 'small, middle and great powers' has a much longer pedigree and was the normal term used in all European languages 'until well into the twentieth century'. In this context, 'small power' emerges simply as a residual category characterised as those powers that were not a 'great power' or a 'middle power'. It therefore necessarily lacks definitional precision. International politics also focused on war and diplomacy between the great powers (later superpowers), to the relative exclusion of small powers, which were deemed as inconsequential and weak. As such, it was not difficult for a confusion of 'weak states' with 'small states' to become established within the realist approach to international politics; but while small states are weak powers they are not necessarily weak states. Following Barry Buzan (1983, pp. 65–69), a small state can be a strong state: strong states have high legitimacy, weak states do not; strong powers have high capabilities, weak powers do not. This distinction is frequently ignored, even in relatively modern literature. For example, Maniruzzaman in his 'The security of small states in the Third World' (1982) writes of 'small states' when he means 'weak powers', while Handel in *Weak States in the International System* (1981) identifies 'weak powers' as 'weak states', following the older usage of the term in which they are synonymous with small states. But small states can be, as argued above, strong states and can use that to advantage in international relations to 'win' in conflicts with much larger states, as is clearly indicated in recent examples from Iceland, Malta and Antigua (Cooper and Shaw, 2009).

It is therefore important to arrive at a definition of the concept of a small state in international politics that does not perpetuate these confusions. Unfortunately it has proved very difficult. In part, it is because the historically dominant realist school of international relations emphasises the capabilities of states as the critical dimension of state power. Vital (1967) is a good example. In his study on *The Inequality of States* (1967, p. 8) a small state is considered as one having '(a) a population of 10–15 million in the case of economically advanced countries; and (b) a population of 20–30 million in the case of underdeveloped countries'. Handel (1981, p. 31), although not precisely defining a small state, explicitly excludes what he terms 'mini states' from his study, these being states with a population below five million and a GNP below US\$1bn. On these measures, following Vital, the vast majority of contemporary states are 'small states', whereas following Handel, nearly half the world's states are close to qualifying as 'mini states', given that Baldacchino (2009, p. 23) has recently shown that the population size of the median jurisdiction of 237 jurisdictions listed in the 2007 edition of the *CIA World Factbook* is Finland, with 5.3 million people. In such circumstances it is easy to see why in traditional international political analysis small states are both overlooked and ill defined—given their number they are 'whatever criterion is adopted ... too broad a category for purposes of analysis' (East, 1973, p. 466).

This, however, is clearly insufficient. In academic terms it distorts the study of international politics by privileging theories based on power while in practical terms it largely ignores the 39 small states (population around one million or less) created by decolonisation since 1960, plus others of longer provenance. It also fails to distinguish sufficiently *between* small states, even within a paradigm framed primarily by a traditional interest in security. Ken Ross (1997, p. 71) provides a useful correction to this situation. Although he accepts a definition of one million population for 'small states' as useful for many purposes, he argues that it is 'too constraining' in 'the arena of international security'. He therefore proposes a threefold distinction: states with a population between one and five million, which he terms 'small states' ($n = 46$); states with a population between 100,000 and one million, which he identifies as 'mini-states' ($n = 27$); and states with a population below 100,000, seen as 'micro-states' ($n = 15$). 'In global security affairs', he argues 'there is a need to appreciate that the group of states with populations between one and five million more readily identify with the "very small" (below one million population) than with larger states' (Ross, 1997, p. 71). In support of this proposition, he lists 32 Commonwealth and other small island states (over all three categories) which, although they could not 'put together a military division', were 'developing a distinctive approach to international security' based on being 'good international citizens' (Ross, 1997, p. 72). In short, in international politics there is a need to 'disaggregate' small states from larger states and to discriminate between them by 'issue area', in this instance 'security'. On this reading the concept of 'small state' in international politics is more elastic than it is in international economics, added to which is yet another category to confuse the picture—'the subnational island jurisdiction'.

In a body of work pioneered in the last 10 years or so, Godfrey Baldacchino (2010) has evolved the concept of the subnational island jurisdiction (SNIJ). In essence, the SNIJ is a distinct political jurisdiction that exercises a large measure of autonomy in a number of 'issue areas', providing it with economic flexibility and a form of 'quasi-sovereignty' (Baldacchino, 2004) that it exercises in its own interest. Examples would be the majority of British Overseas Territories and the Crown Dependencies, the Netherlands Antilles (dismembered into five separate parts on 10 October 2010) and Aruba in the Caribbean, and many territories associated in various ways with sovereign countries throughout the world, nearly all of which have small populations and are islands (Baladacchino and Milne, 2006, Table 1).

The issue here is not, as in international economics, whether islandness is distinct from smallness, but whether the 'jurisdiction' exercised by the SNIJs is 'functionally' equivalent to the sovereignty exercised by 'small states'. In the final analysis it is not, because SNIJs are ultimately accountable to and subordinate to their metropoles, but in many other ways, and in particular in economic 'issue areas' and in various regional intergovernmental organisations, it is. In recent years this has led to the growth of 'para-diplomacy', defined as 'all those external activities by non-sovereign jurisdictions that stimulate and approximate the formal, legal and recognised diplomatic practices of sovereign states' (Bartmann, 2006, p. 544). These now span an ever increasing range of activities associated with the development of 'multi-level governance' to facilitate and regulate globalisation. They can include 'formal' diplomatic relations, as is evident in the Cook Islands, which has separate diplomatic relations with over 20 states at high commission or embassy level, matching and exceeding the level and scope of relations entered into by some small sovereign states in the same region. They also can, and for many years have, involved representation as full (or on occasion associate) members in regional organisations in the Caribbean and the South Pacific, as well as in the UN.

In the light of all these difficulties it is not surprising that the concept of 'small state' in international politics remains imprecise and contested. Further, it is not surprising to find even less discussion of it than in international economics, along with an even greater willingness to avoid any definition, the usual form of words being that any definition is meaningless and that consequentially a small state is whatever the author(s) define it to be. The study of small states in international politics in a scientific manner thus remains undeveloped in spite of the modest but growing literature devoted to them (see Neumann and Gstohl, 2004).

The Small State in the Globalised World

In addition to the academic discussion of small states, a large number of technical and policy studies on small states have emerged in recent years, which have contributed to our understanding of the international political economy of small states. The two major streams of this work, as noted earlier, have been the studies within the UN, first in UNCTAD and then by a scattering of other UN agencies, which have focused on SIDS, and within the Commonwealth, where the focus has been on small states in general. Small states have also featured in studies by the World Bank, the World Trade Organisation (WTO), and negotiations for the Free Trade Area of the Americas (FTAA).[1]

Although some of this work extends back to the early 1980s, the catalyst for the most recent phase of such studies was the publication in 1997 of the Commonwealth Secretariat's report *A Future for Small States: Overcoming Vulnerability*. This updated the 1985 report *Vulnerability: Small States in the Global Society* and extended it by including economic and environmental dimensions as well as the geopolitical and security dimensions, which were the almost exclusive concern of the earlier study. The 1997 report set the study of small states on a new path, promoting the international *economic* interests and vulnerabilities of small states over their international *political* (security) concerns. The expression of this new emphasis within the Commonwealth Secretariat itself was a transfer of lead responsibility for small states within the organisation from the Political Affairs Division to the Economics Affairs Division following the publication of the report.

In retrospect, the 1997 report set the parameters for subsequent considerations of small states in three areas. The first was to establish a 'working definition' of a small state, which was subsequently adopted by the World Bank and others. This defines a small state as one with a population of 1.5 million. In so doing both the Commonwealth Secretariat and the World Bank acknowledged: 'that no definition, whether it be population, geographical size or GDP, is likely to be fully satisfactory. In practice there is a continuum, with states larger than whatever threshold is chosen sharing some or all of the characteristics of smaller countries' (Commonwealth Secretariat/World Bank, 2000, paragraph 8). On a strict interpretation of these criteria there are currently 50 small states with a population of 1.5 million or less (excluding the Vatican), 45 of which can be classed as developing or transition states; but in practice both the Commonwealth Secretariat and the World Bank admit several exceptions drawn from more populous states to be considered as small states, so as noted it is a 'working definition' not a 'definitive' one.

Second, it privileged vulnerability as the key dimension of small states compared with larger states. This was not new because vulnerability was a key theme of earlier studies, but it now sought to give it a *quantitative* as well as a qualitative dimension. The early work was pioneered by UNCTAD and developed and published by Briguglio (1995) as a vulnerability index. In 1996 the Commonwealth Secretariat initiated its own study for an index and, following the publication of the 1997 report, commissioned a study based on the finding that the volatility of output (GDP) is significantly greater for small states than for larger states. It was published in 2000 (Atkins *et al.*, 2000) and confirmed the economic vulnerability of small states. In a sample of 111 developing countries (34 of which had a population of 1.5 million or less and were categorised as small as compared with 77 with a population above this level and considered as large), 26 of the 28 most vulnerable were small states whereas all of the 28 countries with low vulnerability were large. In the category of higher medium vulnerability, six of the 28 were small states and in the category of lower medium vulnerability, two of the 27 states were small (Atkins *et al.*, 2000, Table 7).

Third, it located the main problem for small states as the impact of globalisation on them and in particular their fears of marginalisation in the world economy. A Commonwealth Secretariat/World Bank Joint Task Force was established in 1998, which submitted its final report in March 2000. The report analysed and confirmed the vulnerabilities of small states and then focused in detail on four areas of special relevance to small states: how best to tackle volatility, vulnerability and natural disasters; strengthening capacity; issues of transition to the changing global trade regime; and key challenges and opportunities arising from globalisation. The Development Committee of the World Bank considered the report at its April 2000 meeting and broadly accepted the recommendations, but did not support the creation of a new category for small states that would give them 'special and differential treatment' analogous to that given to the least-developed countries. Those supporting further consideration of such a status included some of the largest states (Brazil, China, India and Russia) and some regional middle powers that acted as spokesmen for small states (Australia and Canada), while those most opposed included many Latin American countries and France. The United Kingdom was non-committal, as was the United States (Sutton, 2001a, pp. 100–101). The failure to get 'special and differential treatment' was a disappointment for some small states and it was left for them to follow this route in other international organisations.

Taking their cue from the report, one of the most promising areas was the decision of the WTO (2001, paragraph 35) in the Doha Ministerial Declaration to establish a

Work Programme on Small Economies. This would 'frame responses to the trade-related issues identified for the fuller integration of small, vulnerable economies into the multilateral trading system' (i.e. address marginalisation) but without '[creating] a sub-category of WTO members'. Most of the work on small states was channelled through 18 'dedicated sessions' of the Committee on Trade and Development held between 2002 and the end of 2008. A major change here was the switch in conceptual terminology from 'small state' to 'small vulnerable economy', carried alongside the continuing problem of definitional imprecision. For example, an early literature review by the WTO surveyed as 'small states' those with a population under five million and, unsurprisingly, found it hard 'to find that "one" aspect of smallness that is essential in the characterisation of small economies and in distinguishing them from large states' (WTO, 2002, paragraph 5). The work of the 18 sessions compounded this imprecision by scrupulously avoiding any formal definition of a 'small vulnerable economy', leaving it to the negotiating groups on agriculture, NAMA, services, fisheries subsidies and trade facilitation, among others, to consider specific measures to address the concerns of small, vulnerable economies as they saw fit. In these sessions some progress was apparently made (Werner, 2008), but the idea of a 'small vulnerable economy' receiving 'special and differential treatment' simply as a 'small vulnerable economy' made little, if any, progress, given the initial prohibition on creating a special subcategory of such economies, which effectively precluded conceptual clarity in identifying such economies for special consideration (Grynberg and Remy, 2003; von Tigerstrom, 2005; Bernal, 2009, pp. 18–20).

A similar outcome is to be found in the FTAA negotiations. These began in 1994 and explicitly recognised a need to accommodate the interests of 'smaller economies' (not small vulnerable economies) in the programme to establish a free trade area. To facilitate the process a Working Group on Smaller Economies was established in 1995, which held a number of meetings to identify the factors that could affect the participation of smaller economies and assist their adjustment to the process, before completing its programme in 1997. In 1998 this 'group' was 'reborn' as the Consultative Group on Smaller Economies, holding 24 meetings until January 2004. Once again there was no attempt to define a 'smaller economy', although in its early years its first chairman, Richard Bernal, presented a paper setting out some of the issues (Bernal, 2000). Consequently, the work on accommodating the interests of 'smaller economies' was left largely to the nine negotiating groups, which were issued with guidelines to take into consideration in their deliberations on differences in the levels of development and size of economies. Again some progress was made, particularly in identifying technical assistance needs for 'smaller economies' and the need for longer periods of adjustment to a free trade area, but as with the WTO process it was brought to a premature end when the whole process 'stalled' as the result of differences among the major countries involved.

What these two examples show is that lack of clarity in definition can itself be an objective in international organisations. The decision, or rather non-decision in the sense of deliberately not raising the issue, of what constitutes a 'small vulnerable economy/smaller economy' is based on the political calculation that any definition would be contested by some countries, as the evidence of disadvantage arising from smallness is equivocal and contradictory. It would almost certainly also be challenged by other larger developing countries that are also disadvantaged and which also seek special consideration in the global economy. In short, advancing the case for 'special and differential treatment' as a compensation for 'smallness' in international trade and other issue areas faces real political problems. The case for 'special and differential treatment' has a long history in numerous preferential arrangements and special derogations, whether it is trade, development assistance, subscription fees for international organisations, and the like, but the question always is: who is eligible and on what grounds? If small states/economies want eligibility *qua* small states/economies then they have to agree a definition for themselves and others and demonstrate disadvantage. As this is problematic it is, of course, politically useful to avoid the issue, or more exactly be imprecise about what constitutes a small state/ economy in order to maximise visibility and support; but then if that is so it comes with a price—that the case for 'special and differential' treatment has to be made over and over again in each and every circumstance and in each and every setting. In such situations small states more often than not demonstrate a real disadvantage of small size in their 'inherent' negotiating weakness in spite of often having a very good 'technical' case for 'special and differential treatment' (Horscroft, 2005). In all, it is a conundrum and a contradiction without easy resolution. In the case of SIDS, and from long experience in dealing with them in UNCTAD, both Hein (2004) and Encontre (2004) argue for the importance of an agreed definition for SIDS to secure 'special and differential treatment' for them. It is a persuasive argument for SIDS[2] but then again it does not include *all* small developing states even if it includes the majority; and so the conundrum begins again.

In the mean time, small states have two 'champions' in the global arena. One is the World Bank, which continues to maintain its interest in them. In 2005 it joined up again with the Commonwealth Secretariat to commission a follow-up report to the 2000 Joint Task Force report. This identified new challenges for small states and painted a mixed picture of the responses to them by both small states and international donors, which again underlined the continuing difficulties small states faced on account of their specific characteristics and the continuing impact of globalisation on them (Briguglio *et al.*, 2005). In addition, the World Bank has hosted the annual Small States Forum. The first was held in Prague in September 2000 and the most recent in Washington, DC in October 2010. The small states invited included all those developing and transition member states with a population of less than 1.5 million plus Botswana, Jamaica, Lesotho and Namibia. The agenda included discussion of and regular reports on progress in implementing the recommendations in the two Commonwealth Secretariat/World Bank reports, plus specially selected themes of particular interest to small states. The Forum is held over one half day and attracts *circa* 100–200 participants. It has been a useful mechanism to keep track of the World Bank's commitment to small states, but as I noted some 10 years ago the record on the promotion of small states outside the Commonwealth has been weak and 'the Commonwealth will need to be vigilant to ensure this meeting is no more than an annual ritual' (Sutton, 2001b, p. 87). Alas, according to reports given to me by some participants to the Forum, that is precisely what it has become.

The other champion is the Commonwealth (Sanders, 2010). For such a comparatively small international organisation it has an impressive work pro- gramme for small states (Vigilance, 2008), but it can be criticised for two reasons. The first is that it is excessively technical. The Commonwealth Secretariat identifies its interest in small states as facilitating integration into the global economy, developing economic resilience and promoting competitiveness. These are worthy aims, but as its own programme on economic resilience has shown, the development problem for small states is multi-dimensional, with environmental, political and social dimensions as well as economic ones (Briguglio *et al.*, 2008). In comparison with the economic dimensions, these others have been neglected.

Second, although small states command a lot of attention in the division of resources within the Commonwealth Secretariat, they no longer command political attention to the same degree. In 1993, a Ministerial Group on Small States (MGSS) was established for the first time at the Commonwealth Heads of Government Meeting (CHOGM) in Cyprus and met separately to consider the special interests of small states. The MGSS met for a sixth and last time in Nigeria in 2003. The final communiqués of the CHOGM always include a section on small states, but the fact there is no longer a dedicated space for them at the CHOGM necessarily limits their voice and impact on the proceedings and in the work programme, which is decided at the CHOGM. This does not mean small states are entirely without a voice in the Commonwealth, and the convening of the first Small States Biennial Conference in London in July 2010 demonstrates a continuing and perhaps even a reinvigorated interest in their concerns. The final communiqué from this conference sets this out at length, but what is apparent from it and the proceedings of the conference is an absence of a well-thought-out political strategy as to how their case can best be promoted internationally (Commonwealth Secretariat, 2010). It is in this area that the Commonwealth has special expertise and where it could do more to shape and lead the agenda on small states in the international system.

In sum, there has been a flurry of policy initiatives and studies on small states in recent years focused on the economic problems besetting them as a result of globalisation. At issue has been whether small states are specially disadvantaged (or even advantaged) in this process. It is not easy to answer and can quickly change according to circumstance, as witness the debilitating impact of the current financial crises of small states, which have left some of them seriously indebted (UNDP, 2010) and exposed their inherent vulnerability in contrast to their more robust performances in earlier years (Ibitoye, 2009; Commonwealth Secretariat, 2010). It is this concept of vulnerability—how to assess it, how to measure it, and how to combat it—that has been the particularly distinct contribution of these policy initiatives and policy studies to international political economy.[3] In the process a considerable body of data and analysis has been generated that has illuminated the special circumstances of the small state in the global economy, but unfortunately none of this has brought us any closer to an agreed concept of a small state, which, by design and accident, remains both vague and contested.

Conclusion

There is no consensus in the literature on international economics and international politics on what is 'small' or how 'small' is to be defined; nor is there in the practice of international organisations creating programmes for small entities. To begin with what are we considering? Is it a small state, a small island developing state, a small sovereign state, a subnational island jurisdiction, a small vulnerable economy, a smaller economy, or what? If they are different, how are they different? Second, what is the most appropriate measure and the most appropriate cut-off point? Taking just one universal measure—population—is 'small' below 10 million people, five million, three million, one-and-a-half million, or one million? And do we need to distinguish within these categories between small, smaller (mini) and smallest (micro)? Third, do we also need to distinguish between issue areas and levels of development? What is seen as 'small' in international trade is not necessarily the same as what is seen as 'small' in international security, and what is regarded as a 'small' developing country may not be the same as a 'small' developed one. In short, where do we start and does the starting point depend on where we sit (and how we frame the problem)?

None of the above questions is easy to resolve. That is why they have so often been studiously, and deliberately, avoided. Should international political economy follow the same course? There is some merit in definitional flexibility, as Maass (2009) has recently argued, but it necessarily comes at a loss of scientific rigour if not of theoretical insight. In the case of international political economy the latter is all that can be offered given the present state of knowledge, though tempered by two considerations. The first is that many of the concerns of international political economy are policy related, so I suggest we borrow from the policy community. The definition of the small state that has generated the greatest volume of scholarship and interest is that advanced by the Commonwealth Secretariat and the World Bank. As such, the paradigmatic small state is one with a population of 1.5 million or below, with perhaps the recognition that at the upper boundary it can include others if a good case is made to do so. The second is that the same policy community has identified vulnerability as the key characteristic of small states, marking them off from other states. Again this is a contested concept, both in measurement and application, but it has proved to be robust enough to withstand all but the most strident of critics and, in the words of Tony Payne (2004, p. 634): 'it is vulnerabilities rather than opportunities ... that come through as the most striking manifestation of the consequences of smallness in global politics'. None of this is to deny that small states can prosper or develop resilience (Cooper and Shaw, 2009), but vulnerability should be seen as the core characteristic of small states in the contemporary international political economy. It sets them apart from most other states and establishes an agenda in many ways unique to their needs.

Notes

1. There is a parallel and in some areas convergent discussion on small states in the work of the various UN development and environment agencies, in particular in the Barbados Programme of Action adopted in 1994, the Mauritius Programme of Action adopted in 2005 and the recent five-year high-level review of the Mauritius programme, which took place in September 2010. Lack of space precludes their consideration in this article.
2. Even here, however, there are difficulties as the UN has not developed an official list of SIDS. For example, a recent paper identifies Cuba, the Dominican Republic, Haiti and Singapore as SIDS. See UNDESA (2010).
3. In recent years vulnerability has often been conceptualised alongside a companion concept of resilience, which has attracted research by Briguglio and others both to measure it and to

determine its key component parts. Again lack of space prevents an examination of this research in this article.

References

Armstrong, H. and Reid, R. (2006) Determinants of economic growth and resilience in small states, in L. Briguglio, G. Cordina and E. Kisanga (Eds), *Building the Economic Resilience of Small States* (Malta and London: University of Malta and Commonwealth Secretariat).

Atkins, J., Mazzi, S. and Easter, C. (2000) *A Commonwealth Vulnerability Index for Developing Countries: The Position of Small States*, Commonwealth Secretariat Economic Papers, No. 40.

Baldacchino, G. (2004) Autonomous but not sovereign? A review of island sub-island nationalism, *Canadian Review of Studies in Nationalism*, 31(1–2), pp. 77–89.

Baldacchino, G. (2009) Thucydides or Kissinger? A critical review of smaller state diplomacy, in A. Cooper and M. Shaw (Eds), *The Diplomacies of Small States: Between Vulnerability and Resilience* (Basingstoke: Palgrave Macmillan).

Baldacchino, G. (2010) *Island Enclaves: Offshoring Strategies, Creative Governance, and Subnational Island Jurisdictions* (Montreal: McGill-Queen's University Press).

Baldacchino, G. and Milne, D. (2006) Exploring sub-national island jurisdictions: an editorial introduction, *The Round Table*, 95(386), pp. 487–502.

Bartmann, B. (2006) In or out: sub-national island jurisdictions and the antechamber of para-diplomacy, *The Round Table*, 95(386), pp. 541–549.

Bernal, R. (2000) Smaller economies in the free trade of the Americas, WTO Seminar on the Smaller Economies in the Multilateral Trading System, 18 October.

Bernal, R. (2009) *Participation of Small Developing Economies in the Governance of the Multilateral Trading System*, CIGI Working Paper, No. 44 (Waterloo, Canada: Centre for International Governance Innovation).

Briguglio, L. (1995) Small island states and their economic vulnerabilities, *World Development*, 23(9), pp. 1,615–1,637.

Briguglio, L., Cordina, G., Farrugia, N. and Vigilance, C. (Eds) (2008) *Small States and the Pillars of Economic Resilience* (Malta and London: University of Malta and Commonwealth Secretariat).

Briguglio, L., Persaud, B. and Stern, R. (2005) *Toward an Outward-oriented Development Strategy for Small State: Issues, Opportunities, and Resilience Building* (A Review of the Small States Agenda Proposed in the Commonwealth/World Bank Joint Task Force Report of April 2000), Final Draft Report, 8 August.

Buzan, B. (1983) *People, States and Fear* (Brighton: Wheatsheaf Books).

Commonwealth Secretariat (1985) *Vulnerability: Small States in the Global Society*, Report of a Commonwealth Consultative Group (London: Commonwealth Secretariat).

Commonwealth Secretariat (1996) *Small States: Economic Review and Basic Statistics*, Vol. 2 (London: Commonwealth Secretariat).

Commonwealth Secretariat (1997) *A Future for Small States: Overcoming Vulnerability*, Report by a Commonwealth Advisory Group (London: Commonwealth Secretariat).

Commonwealth Secretariat (2010) Marlborough House Small States Consensus, *Small States Biennial Conference*, Marlborough House, London, 28–29 July.

Commonwealth Secretariat/World Bank (2000) *Small States: Meeting Challenges in the Global Economy*, Report of the Commonwealth Secretariat/World Bank Joint Task Force on Small States, March (London: Commonwealth Secretariat).

Cooper, A. and Shaw, T. (Eds) (2009) *The Diplomacies of Small States: Between Vulnerability and Resilience* (Basingstoke: Palgrave Macmillan).

Crowards, T. (2002) Defining the category of small states, *Journal of International Development*, 14(2), pp. 143–179.

Demas, W. (1965) *The Economics of Development in Small Countries with Special Reference to the Caribbean* (Montreal: McGill University Press).

Downes, A. (1988) On the statistical measurement of smallness: a principal component measure of country size, *Social and Economic Studies*, 37, pp. 75–96.

Downes, A. and Mamingi, N. (2001) The measurement of size and implications for the survival of small states in the global economy. Paper presented at the *First Annual Conference of the Global Studies Institute*, Pennsylvania, March.

East, M. (1973) Size and foreign policy behaviour: a test of two models, *World Politics*, 25(4).

Encontre, P. (2004) SIDS as a category: adopting criteria would enhance credibility, in United Nations Conference on Trade and Development, *Is a Special Treatment of Small Island Developing States Possible?* (New York and Geneva: United Nations).

Grynberg, R. and Remy, J. Y. (2003) Small vulnerable economy issues and the WTO, in *Small States: Economic Review and Basic Statistics*, Vol. 8 (London: Commonwealth Secretariat).

Handel, M. (1981) *Weak States in the International System* (London: Frank Cass).

Hein, P. (1985) The study of microstates, in E. Domment and P. Hein (Eds), *States, Microstates and Islands* (London: Croom Helm).

Hein, P. (2004) Small island developing states: origin of the category and definition issues, in United Nations Conference on Trade and Development, *Is a Special Treatment of Small Island Developing States Possible?* (New York and Geneva: United Nations).

Horscroft, V. (2005) Small economies and special and differential treatment: strengthening the evidence, countering the fallacies, in *Small States: Economic Review and Basic Statistics*, Vol. 10 (London: Commonwealth Secretariat).

Ibitoye, I. (2009) Small states in the global economic downturn, in *Trade Hot Topics*, Issue 64 (London: Commonwealth Secretariat).

Jalan, B. (1982) Classification of economies by size, in B. Jalan (Ed.), *Problems and Policies in Small States* (London: Croom Helm).

Kuznets, S. (1960) Economic growth of small nations, in E. A. G. Robinson (Ed.), *The Economic Consequences of the Size of Nations* (London: Macmillan).

Maass, M. (2009) The elusive definition of the small state, *International Politics*, 46(1), pp. 65–83.

Maniruzzaman, T. (1982) The security of small states in the Third World, *Canberra Papers on Strategy and Defence 25* (Canberra Strategic and Defence Studies Centre, The Australian National University).

Neumann, I. and Gstohl, S. (2004) *Lilliputians in Gulliver's World: Small States in International Relations*, Working Paper 1-2004, May (Centre for Small State Studies, University of Iceland).

Payne, A. (2004) Small states in the global politics of development, *The Round Table*, 93(376), pp. 623–635.

Ross, K. (1997) The Commonwealth: a leader for the world's small states? *The Round Table* (Small Statehood and the Commonwealth Reconsidered: Presented to Delegations at the Commonwealth Heads of Government Meeting, Edinburgh, October).

Sanders, R. (2010) The Commonwealth as a champion of small states, in J. Mayall (Ed.), *The Contemporary Commonwealth: An Assessment 1965–2009* (London: Routledge), pp. 83–102.

Sutton, P. (2001a) On small states in the global system: some issues for the Caribbean, in R. Ramsaran (Ed.), *Caribbean Survival and the Global Challenge in the 21st Century* (Jamaica: Ian Randle).

Sutton, P. (2001b) Small states and the Commonwealth, *Commonwealth and Comparative Politics*, 39(3), pp. 75–94.

Taylor, C. (1969) Statistical typology of micro-states and territories: towards a definition of a micro-state, in United Nations Institute for Training and Research, *Small States and Territories: Status and Problems* (New York: Arno).

UNCTAD (1983) *Specific Action Related to the Particular Needs and Problems of Land-locked and Island Developing Countries* (UNCTAD TD/279, 28 January).

UNDESA (2010) *Trends in Sustainable Development: Small Island Developing States* (New York: United Nations).

UNDP (2010) *Discussion Paper: Achieving Debt Sustainability and the MDGs in Small Island Developing States* (New York: United Nations).

Vigilance, C. (2008) Small states and the Commonwealth Secretariat, in L. Briguglio, G. Cordina, N. Farrugia and C. Vigilance (Eds), *Small States and the Pillars of Economic Resilience* (Malta and London: University of Malta and Commonwealth Secretariat).

Vital, D. (1967) *The Inequality of States: A Study of the Small Power in International Relations* (Oxford: Clarendon Press).

Von Tigerstrom, B. (2005) Small island developing states and international trade: special challenges in the global partnership for development, *Melbourne Journal of International Law*, 6(2), pp. 402–436.

Werner, H.-P. (2008) Finding answers to the concerns of small, vulnerable economies in the Doha Round, in L. Briguglio, G. Cordina, N. Farrugia and C. Vigilance (Eds), *Small States and the Pillars of Economic Resilience* (Malta and London: University of Malta and Commonwealth Secretariat).

WTO (2001) *Ministerial Declaration*, 14 November.

WTO (2002) *Small Economies: A Literature Review* (Committee on Trade and Development, WT/COMTD/SE/W/4, 23 July).

Negotiating Crisis: The IMF and Disaster Capitalism in Small States

ANDRÉ BROOME

University of Birmingham, Birmingham, UK

ABSTRACT   *How do small states use international organisations to manage the consequences of exogenous shocks? This article examines this question through exploring how small states negotiate with the International Monetary Fund (IMF) for crisis management support during a period of 'disaster capitalism'. Focusing on the case of Iceland, the article argues that while small states can potentially build scale economies in specialist sectors such as banking, the risks inherent in rapid financial expansion greatly increase their vulnerability to external shocks. In such circumstances, small states are likely to struggle to level the playing field in their attempts to negotiate the constraints and opportunities provided by engagement with the IMF during international crises, when they face higher stakes compared with larger economies and have a narrower policy choice set at their disposal.*

Introduction

Disaster capitalism is defined here as an international crisis episode that is not contained to one particular market-based economy or subset of economies. The consequences of such an episode are often associated with a departure from established forms of economic governance, at both the national and international levels. Indeed, the effects of disaster capitalism may be so severe across open economies that they empower decision-makers to sanction the 'exceptional' use of economic policy tools and regulatory measures that mark a significant departure from prevailing economic policy norms. Such emergency policy measures that break with economic orthodoxy are often crucial to a state's ability to mitigate the domestic fallout from global financial crises. For states that must turn to the International Monetary Fund (IMF) for external assistance during extreme financial distress, however, regulatory flexibility is circumscribed by the formal limits on policy activism that are incorporated within the organisation's loan programme conditions.

Owing to these constraints, few states ever willingly exhibit a preference to turn to the IMF for financial support when faced with an economic disaster. Requesting an

IMF loan is more often than not viewed as a stark symbol of policy failure, with most states preferring instead to call in the IMF only once all other available options have been exhausted. Although tying a government's hands via the commitment mechanisms that are associated with IMF loan programmes can sometimes provide a useful scapegoat opportunity for national policy-makers to minimise blame for a country's economic distress (Vreeland, 2003), the costs of requiring IMF assistance can often prove to be severe in terms of economic sovereignty, electoral support, and policy credibility with both domestic and international audiences.

Small states face the same constraints as most other states during a period of disaster capitalism, with the exception of the largest and most powerful economies. However, this article suggests that sovereignty, electoral and credibility costs are higher in the case of small states seeking financial support from the IMF in comparison with larger states when small states have developed scale economies in niche industries that depend on an exogenous growth model for expansion. Small states are defined here primarily on the basis of population (below 1,000,000 population=small state) rather than employing other potential indicators of smallness, such as gross domestic product or geographical territory (see Maass, 2009). One important reason for selecting this definition is that even when small states have large economies or control a large territory, their low population size limits the effectiveness of the range of policy tools they have at their disposal when faced with financial disaster. Such limitations are *relative* rather than absolute. For example, increasing taxes on a workforce of 300,000 to tackle a blowout in external debt obligations caused by a banking crisis in a small state where the financial sector far outstrips the size of the domestic economy can be expected to be relatively less effective (or at least more likely to involve a steeper tax rise) compared with a state that has a workforce of 30,000,000, where the fiscal cost can be spread more lightly across the population through changes in income, consumption, property and corporate tax rates.

In a similar vein, states with smaller populations can find it difficult to build a broad manufacturing base in the absence of economies of scale. Although their small size can sometimes belie scale economies in specialised industries (see, for example, Herbertsson and Zoega, 2003), the significant challenges faced by small states often constrain efforts to diversify their economies in order to mitigate vulnerability to terms of trade shocks (see Briguglio, 1995; Payne, 2008; cf. Easterly and Kraay, 2000; Armstrong and Read, 2002). In addition, even comparatively wealthy small states with high per capita incomes usually lack the consumption power derived from large domestic markets to shape regulatory regimes in the world economy (Drezner, 2007), and are less likely to gain a seat at the table in key international policy forums (with some exceptions; see, for example, Thorhallsson and Wivel, 2006).

Compared with larger countries, therefore, the governments of small states that become dependent on external drivers of growth have less room to manoeuvre with respect to economic policy adjustment during a systemic crisis episode, with the scope for enacting significant changes in fiscal and monetary policy that can effectively combat exogenous shocks without severe domestic contraction more tightly circumscribed. In particular, states with small populations are likely to find it more difficult to stimulate demand to prop up economic growth—or to soften the effects of a recession—during a balance of payments crisis, and especially in a context of significant capital outflows. In short, size matters in the world economy, and the focus in this article is on exploring *how* size matters with respect to a select group of states with small populations that, having found themselves in financial dire straits, may turn to the IMF for emergency loan programmes.

The article proceeds as follows. The next section discusses the dynamics of small state negotiations with the IMF in the context of a global financial crisis, and highlights the relative differences that small states may face in their interactions with the IMF compared with larger states. The following section examines the political economy of the financial crisis in Iceland, an example where a small state rapidly developed a large-scale banking sector that subsequently increased the country's vulnerability to exogenous shocks. The fourth section explores how the dynamics of small state negotiations with the IMF played out during 2008 and 2009 in Iceland, which became the first Western economy to borrow from the IMF for emergency balance of payments support in more than three decades in 2008. The Conclusion sums up the main findings of the article, and argues that when the trade-off for small states achieving specialised scale economies is increased vulnerability, the support of international organisations such as the IMF can still provide a potentially useful mechanism for negotiating a passage through financial storms, although external assistance is likely to come with a high political price tag for national governments.

Small States and Disaster Capitalism

An epoch of disaster capitalism can potentially empower national policy-makers to employ unorthodox measures to mitigate the effects of systemic crises. Whether this incorporates nationalisation, foreign exchange controls, increases in trade tariffs, monetary activism, or fiscal pump-priming, policy measures that might be considered heresy during a period of 'normal politics' in an open economy can quickly become pragmatic responses as part of a crisis management strategy in a period of 'extraordinary politics'—a break with orthodoxy that can potentially receive widespread political support from different sections of a society (cf. Gourevitch, 1984). In contrast to earlier periods of systemic crisis such as the Great Depression of the 1930s, however, in the contemporary era the scope for radical policy interventions in the domestic economy that mark a break with established norms of economic governance is curtailed when states seek to negotiate an IMF loan programme.

Although relative 'smallness' can provide states with significant economic opportunities as well as constraints in comparison with larger states (Mehmet and Tahiroglu, 2002), and in some cases being small may have no substantive impact on a country's comparative macroeconomic performance (Armstrong and Read, 2003), their lesser size does present a distinct set of challenges for small states when it comes to negotiating with the IMF in the midst of an economic crisis. Four main factors that directly or indirectly relate to a country's size can be identified that might impinge upon a government's negotiations with the IMF for financial assistance during a period of economic distress: strategic importance; involvement of supplementary financiers; volume of financing needs; and availability of policy alternatives.

First, the relative strategic importance of a state can be a crucial determinant of how easy it is to access IMF resources. In particular, states that are of strategic importance to major power creditors on the IMF's Executive Board may find it easier to gain an IMF loan in the first place, while having a powerful patron onside can also circumscribe the IMF's ability to enforce rigorously its policy conditions during the lifetime of a loan agreement (Stone, 2004), thereby enabling borrowers to negotiate greater policy flexibility and more room to move within the constraints of IMF conditionality. In contrast to the prevailing conventional wisdom in much of the international political economy literature, however, major power states do not always determine the IMF's lending decisions due to the organisation's own sources of autonomy and independence (Broome, 2010a; see also Clegg, 2010). Although some small states may be of strategic importance to major power creditors, in general it is less likely that the tail will be able to wag the proverbial dog in the case of small states, compared with larger states where major powers may have greater interests at stake.

The second and third factors that may influence a small state's negotiations with the IMF both relate to the size of the organisation's resources compared with the financing needs of small state borrowers. In the recent past, the IMF's own resources have not been sufficient to meet the balance of payments needs of borrowers experiencing financial distress, which has led to an increased role for supplementary financiers. Both public and private lenders have supported large IMF loans with additional funds, and this has consequently given supplementary financiers greater leverage over the design and scope of IMF programmes (Gould, 2006). The importance of supplementary financiers was illustrated recently in 2010 by the role of the EU in the multilateral economic negotiations leading to the bailout package for Greece, and the EU and the UK's financial involvement in the bailout for Ireland. This can sometimes be a case where 'he who pays the piper calls the tune', although in practice multi-level negotiations over the use of IMF resources are seldom this simple (see Broome, 2008).

The financing gaps of small states are less likely to exceed the IMF's available resources, thereby obviating the immediate need for the IMF to involve other actors to provide additional sources of funding. Being able to negotiate a loan programme without requiring the involvement of supplementary financiers strengthens the IMF's negotiating hand, and diminishes the coordination problem associated with securing additional sources of support. Just as the participation of multiple bilateral and multilateral lenders can—potentially—increase the range of opportunities for a borrower to play-off lenders against each other (or will at least constrain the IMF's freedom of action), the lesser importance of supplementary financiers for small state borrowers can be expected to harden the IMF's negotiating line. This dynamic increases the advantages of negotiating an IMF loan relative to the 'best alternative to negotiated agreement' (BATNA) in the case of small states (see Narlikar in the introduction to this issue).

Small states seeking an IMF loan are also likely to need the organisation's resources much more than the IMF needs to lend to them. There are two facets to this dimension of the negotiating problem for small states in their dealings with the IMF. First, while the IMF's traditional income model requires continual rounds of new lending in order to generate revenue from charges and interest payments (Broome, 2010b, p. 45), the size of the financing gaps in small states is unlikely to involve a particularly large commitment of the IMF's resources measured as a proportion of the organisation's overall loan budget. The potential income gain/ resource expenditure for the IMF is therefore relatively less than what would be required to offset similar financing gaps in larger states (measured as a percentage of GDP). Second, although the IMF has a clear bureaucratic interest in extending new loans to its member states, during a systemic financial crisis small states must compete with the needs of larger states for the organisation's attention. Owing to the nature of its core business, the fortunes of the IMF flourish during a period of disaster capitalism. Even though the organisation's available lending resources more than tripled during the 2008–09 global financial crisis (Woods, 2010), a cashed-up IMF still has limited resources in terms of staff time to commit to the intensive process of loan programme design and negotiation in a systemic crisis, especially if small states seek to play hardball in negotiations over loan policy conditions.

Finally, the BATNA options of small states are highly limited—they are likely to have fewer policy alternatives at their disposal than larger states in the absence of IMF support. For the reasons outlined above, both fiscal and monetary policy tools in small states may be a relatively less effective means to cope with financial distress independently from the IMF, whereas an IMF agreement is often a prerequisite for states to gain access to additional sources of external financial support (Broome, 2008, p. 138). Furthermore, many small states tend to have comparatively greater levels of trade openness and to be more tightly integrated into the world economy than larger states (Easterly and Kraay, 2000; Armstrong and Read, 2002), which also increases the urgency of closing financing gaps in the event of a systemic economic crisis at the same time as diminishing the scope for governments to employ protectionist trade measures to combat capital flight or current account crises.

These four factors are still relative—not absolute—constraints on small state negotiations with the IMF. Indeed, for each of the challenges identified the opposite may hold true in individual cases. Small states may sometimes be of great strategic importance to major powers, the size of their loan requests may involve supplementary financiers and may be of a volume that has a significant impact on the IMF's overall loan budget, or they may have a range of policy alternatives at the ready should negotiations with the IMF stall or result in unsatisfactory outcomes. In this respect, while IMF-member state loan negotiations are not simply a zero-sum game, they do sometimes contain many of the characteristics of a zero-sum game. This is illustrated in Figure 1, where the balance of influence in negotiations can potentially swing like a pendulum in either direction based on the presence or absence of these related bargaining variables. More often than not, however, these four factors are likely to constrain the bargaining hand of small states more tightly relative to their potential impact on larger states.

Small States and High Finance: The Case of Iceland

Multilateral economic negotiation games over policy choices between states and the IMF often involve successive iterations in the case of frequent borrowers, which incorporate relational dynamics such as social learning, rhetorical action and normative persuasion that collectively constitute long-running contests over policy efficacy. For states that have not recently borrowed from the IMF—where the primary lines of policy communication and interaction have been limited to the

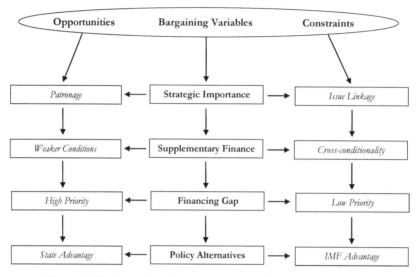

Figure 1. The dynamics of small state negotiations with the IMF.

IMF's regular bilateral surveillance through Article IV consultations (see Broome and Seabrooke, 2007)—the negotiation environment is markedly different.

On the one hand, when traditionally non-borrowing states turn to the IMF for financial support in the midst of an economic disaster, the balance of power may rest with the IMF but the organisation lacks the advantages of familiarity with the policy-making context, and does not have the benefit of previous loan agreement templates with the country to draw on to inform programme design. On the other hand, while actors' existing set of preferences can potentially be reordered—or more radically reconfigured—through iterated negotiation games that take place over a medium-term time horizon, short-term negotiations with the IMF in the absence of a prior track record of interaction and deliberation are less likely to result in lasting changes in policy preferences, assuming the players involved in the game remain the same (cf. Knight, 1995). In this respect, the case of Iceland represents an example of the IMF's engagement with a wealthy small state in severe financial distress in the absence of the analytical shortcuts and policy reform templates that it is common for the IMF to rely on in its negotiations over loan agreements with countries that are more regular borrowers, and where negotiation outcomes are more likely to reflect strategic bargaining, trade-offs and compromises between actors over their existing preferences rather than the reordering or reconfiguration of actors' policy preferences that might be achieved through longer-term processes of social learning and normative persuasion.

Whether states are large or small, previously 'sound' economies can quickly run aground when their growth models rely on financial bubbles that are suddenly punctured by changing fortunes in the world economy. In this respect, Iceland provides an exemplar case of the opportunities and risks that financial integration presents for small states in an era of economic globalisation. Iceland's open economy was described by IMF staff in mid-2008 in glowing terms as 'prosperous and flexible', with 'strong' institutions and policy frameworks and 'enviable' long-term economic prospects (IMF, 2008a). Yet the country's high-debt growth model was premised upon domestic financial expansion funded by a rapid increase in foreign borrowing, equity finance and foreign bank deposits, which exposed the country's financial system to high risks in the event of an exogenous economic shock (Carey, 2009, pp. 8–10).

Iceland's domestic political economy during the 2000s was dominated by an enormous expansion of the banking sector (see House of Commons Treasury Committee, 2009, pp. 8–11), with the combined assets held by financial institutions in 2006 totalling more than eight times the size of the country's gross domestic product, and a domestic credit-to-GDP ratio above 280%. Moreover, the extension of domestic credit in foreign currency-denominated loans as a percentage of total bank credit increased rapidly in the years immediately prior to the financial meltdown of late 2008, with domestic foreign currency loans rising from 43 to 48% of total bank credit between the end of 2005 and the end of 2006 alone (IMF, 2007). One particularly important characteristic of the pre-crisis financial environment in Iceland was the structure of the banking system, which was dominated by three large players: Kaupthing, Glitnir and Landsbanki. While the growth in the international activities of the country's commercial banks offered a source of competitive advantage in boom times, the concentration of the banking system combined with a high-debt/high-growth model funded by foreign credit and savings exposed the economy to significant interest rate and exchange rate risks.

Together, Iceland's three main commercial banks accounted for over seven times the country's GDP and 88% of total domestic financial assets in 2006 (IMF, 2007). The banks' extensive overseas operations, which had expanded at a breakneck pace in a very short space of time, posed a particular set of risks for the stability of the Icelandic financial system in the event of a liquidity crisis. For example, the UK operations of Icesave (a branch of Landsbanki that specialised in retail internet banking) included around 300,000 depositors with total savings of approximately £4.8bn—equal to about two-thirds of the size of Iceland's economy prior to the onset of the financial crisis (Hayes, 2009, p. 1,009).

With the benefit of hindsight, the events of late 2008 illustrate how the achievement of scale economies in small states can have a multiplier effect in terms of a country's vulnerability to exogenous shocks. When the inter-bank lending market seized up in late 2008, Iceland's three major banks were unable to refinance loans through continued foreign borrowing, and all three collapsed within the space of a week during October 2008, prompting the government to assume control of their operations. This contributed to a severe financial crisis for the country as economic conditions worsened across the board, driven by a sudden loss of market confidence with the exhaustion of exchange reserves following the failure of the central bank's efforts to support the value of the króna and the meltdown of the Icelandic stock exchange (Chand, 2009, pp. 2–3). As Figure 2 illustrates, Iceland's economy nosedived in 2008, with a precipitous decline in economic output and domestic demand, large current account and capital account deficits, rising inflation and a rapidly increasing budget deficit. Even more startling, Figure 3 shows the steep increase in the country's external debt between 2003 and 2008, with gross external debt reaching 550% of Iceland's GDP at the end of 2007 (IMF, 2008b).

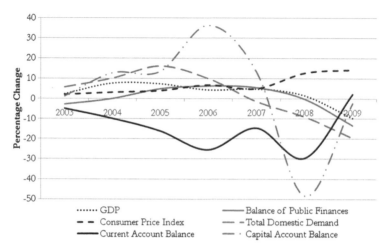

Figure 2. Selected economic indicators for Iceland, 2003–09; 2007 figures are staff estimates; 2008 and 2009 figures are staff projections.
Source: IMF (2009, p. 11).

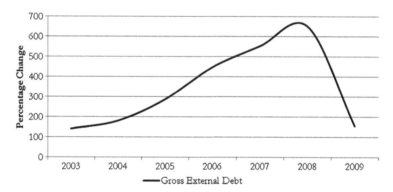

Figure 3. Iceland's external debt, 2003–09; 2007 figures are staff estimates; 2008 and 2009 figures are staff projections.
Source: IMF (2009, p. 11).

Faced with financial dire straits on an unprecedented scale relative to the size of the country's economy, the Icelandic government turned to the IMF for financial support in October 2008 in an attempt to re-establish policy credibility with the international financial community (Jännäri, 2009, p. 21). Following intensive negotiations, this led to a formal request by the government for a relatively large IMF stand-by arrangement loan to the tune of US$2.1bn, amounting to 1,190% of the country's IMF quota (IMF, 2008b). Approval of Iceland's request for financial assistance was expedited under the IMF's Emergency Financing Mechanism procedures in return for swift action by the Icelandic authorities to implement the IMF's crisis management advice, the dynamics of which are discussed in the following section.

The IMF and the Politics of Austerity in Iceland

As Iceland's financial disaster unfolded during late 2008 and 2009, the dynamics of negotiations with the IMF were characterised by a sharply asymmetrical relationship. Applying the framework for small state negotiations with the IMF outlined earlier illustrates the high level of dependence between Icelandic policy-makers and their IMF interlocutors, where most of the bargaining dynamics in negotiations between Iceland and the IMF were in the right-hand column of constraints illustrated in Figure 1 (with the partial exception of the size of the financing gap). For example, although the fate of Iceland's financial system was of great strategic importance to at least one major creditor on the IMF's Executive Board—the United Kingdom—this worked against policy-makers' efforts to negotiate a soft set of loan conditions. Rather than the strategic importance of the country helping to soften the IMF's application and enforcement of policy conditions, the domestic political fallout that stemmed from Iceland's financial crisis in the UK drove British policy-makers to push the IMF to take a hard line to ensure that UK interests were protected in the wake of the Icesave failure (Duncan, 2008; Osborne, 2010; see also House of Commons Treasury Committee, 2009, pp. 19–23). This highlights the extra set of constraints that are placed on small states negotiating with the IMF when the organisation proves useful to a major power creditor for mediating an international economic dispute through linking movement on controversial issues to the approval of IMF loans, such as the efforts to resolve the furore over the freezing of British and Dutch citizens' deposits in Icesave accounts.

In addition, owing to the scale of Iceland's crisis and the unfolding global financial turbulence, the IMF was the principal source of external financial support available for the country to draw upon when the crisis struck, despite efforts by officials from the Central Bank of Iceland to gain access to alternative sources of bilateral credit, including an attempt to negotiate a loan from the Russian Federation in mid-October 2008 prior to the conclusion of negotiations with the IMF (Central Bank of Iceland, 2008a). Further sources of supplementary finance and precautionary credit arrangements that were subsequently agreed between Iceland and Poland as well as the Nordic countries were linked to the implementation of the policy targets detailed in the IMF programme, with a €300m first tranche of a loan from the Nordic countries only becoming available in December 2009 (Central Bank of Iceland, 2009a). This use of cross-conditionality enhanced the IMF's freedom of action and consequently strengthened its negotiating hand in strategic games over loan conditions with Iceland's policy-makers.

While Iceland faced an unprecedented financial crisis relative to the size of its domestic economy (with the level of requested IMF funds equalling approximately 42% of the government's financing gap for 2008–10), the level of priority attached to concluding a loan arrangement with the country by the IMF is harder to gauge. On the one hand, the IMF was all but ignored in the early phase of the credit crunch, with bilateral policy coordination among major economies substituting for the multilateral coordinating function the IMF has historically provided. For this reason, demonstrating its ongoing relevance through concluding a loan agreement with the first Western economy to seek to borrow from the IMF in more than three decades held significant symbolic value for the organisation (see Broome, 2010b). On the other hand, even before the IMF's financial resources were boosted by the Group of Twenty to more than US$750bn, the size of Iceland's stand-by arrangement represented a moderate sum compared with the IMF's overall lending capacity.

With its economy in total meltdown, as illustrated by Figures 2 and 3, Icelandic policy-makers had few viable alternatives to maintaining close cooperation with the IMF in order to avoid a total collapse of the currency and the risk of sovereign default on external debt obligations. One example of the impact that negotiations with the IMF had on the range of options in Iceland's policy choice set is the rapid reversal of a decision by the Board of Governors of the central bank to lower interest rates by 3.5 percentage points to 12% on 15 October 2008, with the next decision over interest rate changes scheduled for 6 November (Central Bank of Iceland, 2008b). In response to pressure from the IMF, this was quickly reversed and the central bank's policy rate was lifted to 18% on 28 October to fulfil one of the IMF's key conditions before a loan request would be put to the organisation's Executive Board (IMF, 2008b). Moreover, while the IMF has exhibited greater tolerance towards exchange rate intervention and current account and capital account restrictions in Iceland in comparison with its policy preferences in previous crisis episodes (cf. Chwieroth, 2010), the organisation successfully pushed for the formalisation of capital controls through the passage of legislation to amend the Foreign Exchange Act of 1992. This formalisation of Iceland's exchange controls (which included the liberalisation of most restrictions on current account transactions) was linked by the IMF to a public commitment by national policy-makers that continuing capital controls would be temporary, with a timetable subsequently established for the gradual removal of all restrictions (IMF, 2009; Central Bank of Iceland, 2009b, 2009c).

With respect to the potential domestic costs of the IMF's financial assistance in terms of economic sovereignty, electoral support and policy credibility, the consequences for Iceland's policy-makers have been severe. The country's economic sovereignty—already highly limited prior to the banking crisis—has been further compromised in an effort to stave off an even more painful process of economic adjustment that might have ensued in the absence of external support to reduce the government's short-term financing gap. Given the poor state of the government's policy credibility with external audiences following the events of September and October 2008, however, the IMF's involvement has the potential to repair some of the reputational damage national policy-makers have suffered, with global credit rating agencies such as Fitch Ratings and Moody's Investors Service explicitly linking future judgements about Iceland's sovereign creditworthiness with the government's strict adherence to the IMF programme (IceNews, 2009).

As a result of the continuing dire state of the economy, Iceland's politicians have borne a steep electoral cost for the crisis (and arguably for bringing in the IMF). For example, in January 2009 the coalition government of Prime Minister Geir Haarde was forced to resign against the backdrop of large public protests aimed at the government's failure to accept responsibility for the causes and management of the crisis. As the final nail in the coffin for the right-wing Independence Party that had ruled Iceland for 18 years, and which had previously been the dominant force in Icelandic party politics for 70 years, subsequent elections in April 2009 handed victory to the Social-Democratic Alliance and the Left-Green Movement, with a new government formed under Prime Minister Jóhanna Sigurðardóttir after the Independence Party received its lowest ever share of the vote (Traynor, 2009). As the case of Iceland illustrates, the politics of austerity comes with a severe electoral cost, regardless of the prospective returns that might be gained from rebuilding a country's policy credibility with international financial markets and the IMF.

Conclusion

The pursuit of scale economies via financial globalisation poses great risks as well as opportunities for small states. The expansion of Iceland's banks into the realm of high finance—previously a key engine of growth for the economy—proved to be the country's undoing. Following the global financial crisis of 2008–09, Iceland has been transformed from a small state model of the advantages of achieving scale through financial integration to a cautionary example of the multiplier effects that financial expansion can generate in terms of vulnerability to exogenous shocks.

As this article has shown, in their negotiations with the IMF Icelandic policy-makers found the deck heavily stacked against them. With the absence of key intervening variables that might have strengthened their negotiating hand (the left-hand column of opportunities in Figure 1), Iceland's political leaders attempted to engage in a scapegoating exercise in order to avoid blame for the country's financial troubles, which spectacularly failed to convince the public, resulting in the government's resignation and subsequent defeat at the polls in April 2009. The case of Iceland holds three important lessons for other small states seeking to negotiate crises and manage a period of disaster capitalism with assistance from the IMF. First, and most obvious, the IMF's involvement may come with a high price in terms of economic sovereignty and electoral support, while agreeing to the IMF's preferences in order to improve policy credibility with external audiences requires a strong commitment to achieving these targets even in harsh economic circum-stances in order to be successful. Second, conducting parallel negotiations with potential supplementary financiers may increase—rather than decrease—a state's dependence on the IMF, owing to the common practice whereby other states may make bilateral credit conditional on adherence to the goals of an IMF programme. Third, emergency policy measures that generate conflict with the interests of major power creditors on the IMF's Executive Board are likely to backfire if larger states are able to link the resolution of bilateral economic disputes to the release of IMF funds.

A final point to conclude with is to note that if the IMF had not been available to act as a lender of last resort to fill the immediate financing gap in Iceland in late 2008, the largest financial crisis in history (relative to the size of the country's economy) would have been likely to result in even greater negative consequences for the country. This would have had serious effects for the economies of other states that experienced the aftershocks of Iceland's financial meltdown, such as the UK and the Netherlands. While the IMF is a far from perfect organ of global economic governance and international crisis management, exploring the counterfactual scenario of what might happen to small states experiencing severe financial distress in the absence of the IMF is therefore a fruitful topic for further research.

31

Acknowledgements

The author thanks the participants at both the 'Small States in the International Political Economy' Workshop at the University of Cambridge in November 2009 and the 'Global Governance in Crisis' Workshop at the University of Birmingham in May 2010 for their helpful feedback and comments on earlier drafts of this article. In addition, the author is immensely grateful to David G. Mayes for his insightful comments and suggestions.

References

Armstrong, H. W. and Read, R. (2002) The phantom of liberty? Economic growth and the vulnerability of small states, *Journal of International Development*, 14(4), pp. 435–458.

Armstrong, H. W. and Read, R. (2003) The determinants of economic growth in small states, *The Round Table*, 92(368), pp. 99–124.

Briguglio, L. (1995) Small island developing states and their economic vulnerabilities, *World Development*, 23(9), pp. 1,615–1,632.

Broome, A. (2008) The importance of being earnest: the IMF as a reputational intermediary, *New Political Economy*, 13(2), pp. 125–151.

Broome, A. (2010a) *The Currency of Power: The IMF and Monetary Reform in Central Asia* (Basingstoke: Palgrave Macmillan).

Broome, A. (2010b) The International Monetary Fund, crisis management and the credit crunch, *Australian Journal of International Affairs*, 64(1), pp. 37–54.

Broome, A. and Seabrooke, L. (2007) Seeing like the IMF: institutional change in small open economies, *Review of International Political Economy*, 14(4), pp. 576–601.

Carey, D. (2009) *Iceland: The Financial and Economic Crisis*, OECD Economics Department Working Papers 725, www.oecd-ilibrary.org/economics/iceland-the-financial-and-economic-crisis_221071065826, accessed 25 March 2010.

Central Bank of Iceland (2008a) Joint press release from the Central Bank of Iceland and the Russian Ministry of Finance, 15 October, www.sedlabanki.is/?PageID=287&NewsID=1917, accessed 25 March 2010.

Central Bank of Iceland (2008b) Board of Governors of the Central Bank of Iceland decides 3.5% policy rate reduction, 15 October, www.sedlabanki.is/?PageID=287&NewsID=1911, accessed 25 March 2010.

Central Bank of Iceland (2009a) First tranche of Nordic loan disbursed to Iceland, 21 December, www.sedlabanki.is/lisalib/getfile.aspx?itemid=7526, accessed 25 March 2010.

Central Bank of Iceland (2009b) Capital account liberalisation strategy, 5 August, www.sedlabanki.is/?PageID=287&NewsID=2196, accessed 25 March 2010.

Central Bank of Iceland (2009c) First stage of capital account liberalisation, 31 October, www.sedlabanki.is/?PageID=287&NewsID=2277, accessed 25 March 2010.

Chand, S. K. (2009) *The IMF, the Credit Crunch, and Iceland: A New Fiscal Saga?*, Working Paper Series 3/09, Centre for Monetary Economics, BI Norwegian School of Management.

Chwieroth, J. M. (2010) *Capital Ideas: The IMF and the Rise of Financial Liberalization* (Princeton, NJ: Princeton University Press).

Clegg, L. (2010) In the loop: multilevel feedback and the politics of change at the IMF and World Bank, *Journal of International Relations and Development*, 13(1), pp. 59–84.

Drezner, D. W. (2007) *All Politics is Global: Explaining International Regulatory Regimes* (Princeton, NJ: Princeton University Press).

Duncan, G. (2008) IMF bailout of Iceland is delayed until fate of UK savers' frozen cash is resolved, *The Times*, 24 October, http://business.timesonline.co.uk/tol/business/economics/article5004002.ece, accessed 25 March 2010.

Easterly, W. and Kraay, A. (2000) Small states, small problems? Income, growth, and volatility in small states, *World Development*, 28(11), pp. 2,013–2,027.

Gould, E. C. (2006) *Money Talks: The International Monetary Fund, Conditionality, and Supplementary Financiers* (Stanford, CA: Stanford University Press).

Gourevitch, P. A. (1984) Breaking with orthodoxy: the politics of economic policy responses to the depression of the 1930s, *International Organization*, 38(1), pp. 95–129.

Hayes, D. G. (2009) Did recent experience of a financial crisis help in coping with the current financial turmoil? The case of the Nordic countries, *Journal of Common Market Studies*, 47(5), pp. 977–1,015.

Herbertsson, T. T. and Zoega, G. (2003) *A Microstate with Scale Economies: The Case of Iceland*, Working Paper 1-2003, Centre for Small State Studies, Institute of International Affairs, University of Iceland.

House of Commons Treasury Committee (2009) *Banking Crisis: The Impact of the Failure of the Icelandic Banks*, Fifth Report of the Session 2008–09, HC 402, www.publications.parliament.uk/pa/cm200809/cmselect/cmtreasy/402/402.pdf, accessed 25 March 2010.

IceNews (2009) Fitch warns Iceland over stability, 29 January, www.icenews.is/index.php/2009/01/29/fitch-warns-iceland-over-stability, accessed 25 March 2010.

IMF (2007) *Iceland: Selected Issues*, IMF Country Report, No. 07/296, August, www.imf.org/external/pubs/ft/scr/2007/cr07296.pdf, accessed 25 March 2010.

IMF (2008a) Iceland—2008 Article IV Consultation Concluding Statement, 4 July, www.imf.org/external/np/ms/2008/070408.htm, accessed 25 March 2010.

IMF (2008b) *Iceland: Request for Stand-by Arrangement*, 4 July, IMF Country Report, No. 08/362, November, www.imf.org/external/pubs/ft/scr/2008/cr08362.pdf, accessed 25 March 2010.

IMF (2009) *Iceland: Stand-by Arrangement—Interim Review under the Emergency Financing Mechanism*, IMF Country Report, No. 09/52, February, www.imf.org/external/pubs/ft/scr/2009/cr0952.pdf, accessed 25 March 2010.

Jännäri, K. (2009) Report on Banking Regulation and Supervision in Iceland: Past Present and Future, Reykjavik, Prime Minister's Office, http://eng.forsaetisraduneyti.is/media/frettir/KaarloJannari__2009.pdf, accessed 25 March 2010.

Knight, J. (1995) Models, interpretations, and theories: constructing explanations of institutional emergence and change, in J. Knight and I. Sened (Eds), *Explaining Social Institutions* (Ann Arbor, MI: University of Michigan Press), pp. 94–119.

Maass, M. (2009) The elusive definition of the small state, *International Politics*, 46(1), pp. 65–83.

Mehmet, O. and Tahiroglu, M. (2002) Growth and equity in microstates: does size matter in development? *International Journal of Social Economics*, 29(1/2), pp. 152–162.

Osborne, A. (2010) Fresh Icesave hitch puts Iceland IMF rescue in doubt, *The Daily Telegraph*, 13 January, www.telegraph.co.uk/finance/newsbysector/banksandfinance/6982676/Fresh-Icesave-hitch-puts-Iceland-IMF-rescue-in-doubt.html, accessed 25 March 2010.

Payne, A. (2008) After bananas: the IMF and the politics of stabilisation and diversification in Dominica, *Bulletin of Latin American Research*, 27(3), pp. 317–332.

Stone, R. W. (2004) The political economy of IMF lending in Africa, *American Political Science Review*, 98(4), pp. 577–591.

Thorhallsson, B. and Wivel, A. (2006) Small states in the European Union: what do we know and what would we like to know? *Cambridge Review of International Affairs*, 19(4), pp. 651–668.

Traynor, I. (2009) Icelandic caretaker government wins general election, *The Guardian*, 26 April, www.guardian.co.uk/world/2009/apr/26/iceland-election-government, accessed 25 March 2010.

Vreeland, J. R. (2003) Why do governments and the IMF enter into agreements? Statistically selected cases, *International Political Science Review*, 24(3), pp. 321–343.

Woods, N. (2010) Global governance after the financial crisis: a new multilateralism or the last gasp of the great powers, *Global Policy*, 1(1), pp. 51–63.

Looking for Plan B: What Next for Island Hosts of Offshore Finance?

MARK P. HAMPTON* AND JOHN CHRISTENSEN**

*Kent Business School, University of Kent, Canterbury, UK

**Tax Justice Network, London, UK

ABSTRACT   *This paper examines offshore finance centres and tax havens that are hosted by small island economies (SIEs). In many cases, hosting offshore finance has been a lucrative activity for SIEs since the 1960s in terms of employment (direct and indirect) and overall contribution to GDP and government revenues. Despite the scale and reach of the global offshore economy, at present many SIE hosts face an unsettled future in light of significant international pressure from nation states, international organisations such as the EU and OECD and, increasingly, from civil society in both the developed and less-developed world. Given the economic importance of hosting offshore finance for many SIEs around the world, the development options facing many island jurisdictions are discussed. The paper poses the fundamental question: what has changed since the major initiatives around the year 2000? Then the situation facing many SIE hosts, the changing global political economy and their shifting negotiations and alliances within it are discussed.*

Introduction

Facing sustained pressure from the US Obama Administration, the Organisation for Economic Cooperation and Development (OECD) and the European Union, the Swiss authorities look likely to abandon or significantly modify their banking secrecy laws in the near future (*International Herald Tribune*, 2009). Other tax havens are coming under similar pressure to modify their laws and treaty arrangements to improve international cooperation on tax information-sharing. In January 2010 the OECD announced that tax havens had signed more than 300 tax information exchange agreements in the preceding 12 months, signalling what they described as 'one of the big success stories of the G20' (Houlder, 2010a, p. 1). The European Court of Justice has also taken a lead in tackling harmful tax practices, and its landmark judgment of April 2005, the Halifax Case, ruled against transactions that have tax avoidance as their sole purpose.[1] Additionally, the British, French and other OECD governments are supporting new accounting rules that will require multinational companies to publish accounts for all subsidiaries in the countries where they operate; rules designed to reduce radically the scope for shifting profits to tax havens (Houlder, 2010a, p. 2).

Cumulatively these pressures seem likely to have a significant impact on small island economies hosting offshore financial centres (OFCs), some of which were already facing budgetary crises arising from other factors. For example, in September 2009 the government of the Cayman Islands, a British Overseas Territory, sought emergency backing from the UK Foreign and Commonwealth Office to borrow £278m on the private markets. In response, a Foreign Office minister proposed radical reform of their tax arrangements, stating: 'I fear you will have no choice but to consider new taxes, perhaps payroll and property taxes such as in the British Virgin Islands' (Mathiason, 2009, p. 7). In October 2009, UK Treasury officials advised senior politicians from the British Crown Dependencies that the EU Code of Conduct Group on Business Taxation would not accept their proposed tax reforms, placing Guernsey and Jersey in potential fiscal crisis (Quérée, 2009). In February 2010, in response to a statement from the UK Secretary to the Treasury about multinational company reporting requirements,[2] the business correspondent of the *Jersey Evening Post* commented: 'It's not only banking secrecy that's dead. Tax avoidance—the industry that helped Jersey become so prosperous—now appears to be on its last legs as well', concluding that 'it would be a good idea to start planning now for this brave new world' (Body, 2010).

Unlike the multilateral initiative in the 1990s against tax havens, spearheaded by the OECD, which fizzled out when the US Bush Administration withdrew its support in 2001 (Palan *et al.*, 2010, p. 224), the current initiatives are being driven by three power blocs, the Group of 20 countries, the EU and the US, all of which are confronted by the deepest recessions experienced since the 1930s. Domestic budgetary pressures within these economies have played an important part in swinging the pendulum away from tax havens. In this context, preparing development strategies that reduce dependence on rent incomes from what the OECD terms 'the tax industry' (OECD, 1998) should be regarded as a priority for small islands.

Theorising Island Hosts: Path Dependence

Although there is a growing literature on small island economies (SIEs),[3] this paper focuses on one aspect, path dependence, as an analytical tool to examine SIE hosts of offshore finance. Scott (2001, p. 367) defines path dependence as 'systems in which early choices or actions, often determined by transient conditions, bias subsequent development in favour of particular outcomes'. Martin and Sunley (2006, p. 402) add, 'The past thus sets the possibilities while the present controls what possibility is to be explored'.

For small islands that are former colonies, path dependence was applied by Acemoglu *et al.* (2002), who discussed the 'reversal of fortune' concept; however, they were criticised by Austin (2008) for over-simplifying the lines of causation over time. Feyrer and Sacerdote (2009) examined the links between colonialism and the modern-day income of islands, pointing to an overall positive correlation between islands being colonial possessions and present national income levels. However, Bertram (2007) argues that the type of approach used (econometrics) had key flaws in its data-gathering and demonstrated a limited understanding of the specifics of SIEs.

MacKinnon *et al.* (2007) argue that when analysing a particular phase in a region's development it is crucial also to consider the existing social relations that shape localised practices and operational routines. This requires a focus on specific circumstances that can give rise to 'lock-in' to a particular form of development in which local economic actors pursue a course of action that may ultimately lead to an 'economic cul-de-sac'. This paper follows this line of logic, arguing that actions and decisions at certain times, in this case to allow OFC firms to 'locate' in SIEs, are likely to have 'locked-in' these jurisdictions to a development path that may indeed prove to be what MacKinnon *et al.* (2007) dub an 'economic cul-de-sac'.

Here this notion is developed further, and it is argued that events since 2007 and 2008, including exogenous shocks, have exacerbated this trend. According to Grabher (1993) and Hassink (2005) there are three distinct aspects of 'lock-in': functional, cognitive and political. Functional lock-in arises from the web of relations between firms, forms of production or economic activity, and the links created between customers and suppliers. Cognitive lock-in (or world view) arises from failures to interpret correctly signs of external change, and the failure of collective learning mechanisms. Finally, political lock-in refers to social relations and the power dynamics that underpin economic development, especially failures of political, business and labour leaders to adapt policy mechanisms to allow local innovation and learning. Of these three aspects, Hassink (2005) argues that cognitive and political lock-ins are closely related and, significantly, can cause outcomes that 'paralyse competition and tranquillize large industries' (Hassink, 2005, p. 523). An example of this can be found in the case of Swiss private banks, which, faced with mounting pressure to abolish the protective mechanism provided by that country's banking secrecy law, feel themselves unable to compete effectively at global level (Zaki, 2010). Here, we would agree with MacKinnon *et al.* (2007, p. 5), who comment that the net effect of these aspects of lock-in further leads to 'collective myopia', that is, a legacy of inherited social infrastructures that actively discourages innovation and new business start-up. Arguably this can be observed in Jersey and many other SIE hosts of offshore finance.

Within the path dependence literature, a link can also be established to *place* dependence (Martin and Sunley, 2006), as despite the so-called 'end of geography' notions (O'Brien, 1992), physical location, by which we mean proximity to major financial centres such as London or New York, remains a key attraction for the largest OFCs (Hampton, 1996a; Hampton and Christensen, 2007; Markoff, 2009).

Hampton and Christensen (2002, p. 168) argued that SIE hosts 'have become locked into their relationships with the offshore finance industry by their dependence upon the earnings potential of predominantly imported skills and expertise, and their lack of skills and knowledge in alternative sectors. This means that any attempts at diversification into other sectors would be constrained by the need for wholesale re-skilling and the acquisition of new knowledge bases.'

Contrary to Park's argument (1982), the experiences of SIEs such as the Channel Islands, Cayman, and Turks and Caicos, demonstrate that hosting OFCs does not lead to transferable knowledge gains or increasing entrepreneurial flair because the majority of the activities located in these OFCs consists of what could be termed 'wrapper' activity, that is, creating and administering structures that give the form but not the substance of a functional presence. Even where these islands have created niche markets for themselves (e.g. securitisation in Jersey, captive insurance in Guernsey), most of the specialists working in these fields have been imported from elsewhere.

The Emergence of Offshore Finance

Offshore finance's rise since the 1960s[4] has been detailed by Hampton (1996a), Palan (2003), Palan *et al.* (2010) and Sharman (2006), but here we can note certain key points. First, there has been the instrumental role played by OECD states (particularly the UK) in creating and supporting tax havens and offshore finance[5] in many small jurisdictions, especially former colonies and continuing dependencies such as the Cayman Islands in the 1960s and Vanuatu in the 1970s.

Second, this explicit support from mainland states, combined with the expansion of globalising financial capital (initially US and Canadian banks plus UK merchant banks), met the rising international demand for both specialist retail banking (asset management for wealthy individuals) and wholesale banking (Eurocurrency and Eurobond markets) for other banks and multinational corporations.

Third, such initiatives overlapped with the personal interests of key local actors such as the lawyers Vassel Johnson in Cayman and Reg Jeune in Jersey, the latter's practice going on to become a prominent member of the 'Offshore Magic Circle'[6] of law firms, with Jeune taking a leading political role as President of the States of Jersey's Policy and Resources Committee. These pioneer law firms were able to make sizeable profits in the new world of offshore finance as it grew in the island hosts. By the 1990s the so-called shadow economy of offshore finance was host to an estimated US$11.5 trillion of assets belonging to wealthy individuals (Tax Justice Network (TJN), 2005), and including differing types of financial structure (varieties of foundations, trusts, tax-exempt companies, international business companies, offshore funds, hedge funds, structured investment vehicles, captive insurance, etc.).

After two decades of benign neglect by the international community, however, and despite *ad hoc* and relatively small-scale initiatives from individual countries' revenue authorities to combat the most blatant tax dodging, by the mid-1990s the international context within which tax havens operate was beginning to change. Initially the islands were able to resist most of the pressures for change. The 1998 Harmful Tax Competition initiative (OECD, 1998)—launched by the OECD at the behest of the G7—was strongly opposed by the 35 tax havens, mostly SIEs, subsequently blacklisted in 2000 (Sanders, 2002; Vlcek, 2007). Their opposition was supported by the incoming George W. Bush administration, whose Treasury Secretary, Paul O'Neill, effectively ended the initiative when he withdrew US support in May 2001 (Palan *et al.*, 2010, p. 217). Lacking the necessary political support, the OECD persevered through its Global Forum process with negotiating agreements with so-called 'cooperating jurisdictions' to remove some of their harmful tax practices. Other changes were also underway, for example the European Court of Justice ruling mentioned above, but political agreement on the need to tackle banking secrecy and strengthen international cooperation with tax information exchange was effectively stalled until April 2009, when the G20 nations, led by British Prime Minister Brown, French President Sarkozy, German Chancellor Merkel and US President Obama, launched a new initiative against tax havens, with the OECD as the prime agent (*La Tribune*, 2009).

The 2009 initiative differed in several respects from the 1998 Harmful Tax Competition programme. First, the OECD targeted a broader range of tax havens, including OECD nations such as Austria, Luxembourg and Switzerland. Crucially, by this stage amendments to Article 26 of the OECD's Model Agreement for Tax Information Exchange included a facility for overriding banking secrecy laws where evidence existed of criminal activity, including tax evasion. Second, the OECD made a distinction between cooperating jurisdictions, specifically those that had already negotiated a minimum of 12 tax information exchange agreements (the so-called White List jurisdictions), from those that had agreed to cooperate but had not achieved the minimum threshold (Grey List jurisdictions), and those that still resisted cooperation (Black List jurisdictions). Third, the public mood had also swung against tax havens, with civil society coalitions in Europe, North America and various developing countries calling for measures to close the loopholes exploited by those who use tax havens.[7] In August 2009 OECD Secretary-General Angel Curria felt sufficiently optimistic to state: 'It seems almost unbelievable, but the era of banking secrecy for tax purposes will soon be over. In tomorrow's world, there will be no more havens in which to hide funds from the taxman' (Gurria, 2009).

Critics such as the Tax Justice Network, however, have argued that the OECD initiative lacks coherence because it fails to target main players such as the UK and US (*Le Monde*, 2009). The TJN, a global civil society coalition that coordinates advocacy efforts for international cooperation on tax matters, published an alternative list of 'secrecy jurisdictions' in November 2009 that ranked the US at the top, and included the UK in the top five.[8] Critics have also argued that the bilateral tax information exchange agreements at the heart of the OECD programme are weak and ineffective at deterring the use of tax havens as centres for tax evasion (*Financial Times*, 2009). As an alternative they propose the automatic information exchange model adopted by the European Union in 2005 (Global Financial Integrity *et al.*, 2009), which they claim has a stronger deterrent effect than the 'upon request' system used in the OECD Model Agreement. Importantly, the jurisdiction of the EU Savings Tax Directive (STD) extends beyond the member states to include their dependent territories such as the British Crown Dependencies and the Dutch Antilles.

With strong political pressure from France and Germany (*La Tribune*, 2009) and from powerful coalitions within the European Parliament, in early 2010 the European Commission tabled proposals that, once implemented, will strengthen the STD and extend its scope.[9] Specifically, the revised STD includes provisions requiring information exchange for trusts, foundations, shell companies and other legal persons. This amendment radically transforms the activities of many SIE tax havens, which use Anglo-Saxon trust laws and nominee company directors as the principal basis for shielding clients from external investigation.

The strengthened STD is likely to be agreed in 2010 and implemented by 2012. The revised model will probably become the global standard for international information exchange, with a number of Latin American countries indicating that they would like an opportunity to pilot this standard with EU support.[10] In addition to strengthening its STD, the EU's Code of Conduct Group on Business Taxation has required the removal of tax measures deemed harmful to the Single Market. The powers of this Group have also extended to dependent territories, some of which have been required to alter radically their tax regimes to remove preferential treatments targeted at companies registered but not trading in tax havens. The British Crown Dependencies, for example, have had their domestic corporate tax regimes overturned on the basis that they 'ring-fence' preferential treatments from local investors (Houlder, 2009). EU intervention has forced Guernsey, the Isle of Man and Jersey to transform their tax regimes, with the latter facing significant structural budget deficits as it struggles to finance public spending while also retaining some attractions as a tax haven.

Unlike in 2000, tax havens are now finding it difficult to resist international pressure for reform (Palan *et al.*, 2010). This is partly because some of them, the Cayman Islands and Jersey, for example, themselves face severe budgetary pressures. As noted earlier, the former has been required by the UK government to introduce new taxes to overcome a severe structural budget deficit (Mathiason, 2009). Attempts to coordinate a counter-attack against anti-tax haven measures have been inhibited by strong campaigns in France, Germany and the UK sparked by revelations of how banks operating from Liechtenstein and Switzerland have actively supported tax evasion by wealthy clients.[11] In the US, public opinion was mobilised during the 2008 presidential campaigns by Barack Obama's regular references to an office block in Grand Cayman that houses over 12,000 registered businesses: 'either this is the largest building in the world or the largest tax scam in the world. And I think the American people know which it is. It's the kind of tax scam that we need to end' (Evans, 2009).

It is unlikely that the initiatives set in motion in 2009 will cause the demise of all tax havens; but there is little doubt that the combination of measures by the OECD and EU will restrict the activities of existing actors and radically reduce their ability to resist requests for international cooperation in tackling tax evasion. As a result of international pressure, banking secrecy laws will probably be degraded to the point of being disabled within the course of the coming decade (*Financial Times*, 2009), and calls are being made for trusts and shell companies to be made more transparent (Global Financial Integrity *et al.*, 2009). Cumulatively this will reduce the scope of many SIE tax havens that rely heavily on secrecy as their prime attraction.

It is not clear whether the islands will be able resist the current wave of initiatives against tax evasion as they did in the early 2000s. Their successful counter-attack against the OECD Harmful Tax Competition initiative used accusations of 'fiscal colonialism' against the OECD countries, and portrayed the battle as a 'brave fight' of the small and 'powerless' against the 'cartel' of OECD countries (Mitchell, 2001; Sanders, 2002). More recently, accusations have been made about a 'witch-hunt' against Caribbean tax havens (Hutchinson-Jafar, 2009), but the inclusion of Austria, Luxembourg and Switzerland on the 2009 Grey List undermines the earlier arguments about OECD tax havens 'ganging-up' against their SIE rivals, and additionally, many of the functional SIE tax havens wish to shed their image as bolt-holes for tax evaders and consequently do not want to be seen to be publicly resisting pressures to negotiate tax information exchange agreements with third-party countries.

What Happens Next?

In 2005, Richard J. Hay, a London-based lawyer acting as adviser to SIE tax havens, strongly advised his clients to support the OECD's 'on request' model for information exchange, as the alternative, the EU's automatic exchange process, would significantly affect demand for offshore private wealth management (Hay, 2005). OECD countries have sought to establish the on-request model as the global standard, and the majority of tax havens listed on the OECD Grey List in April 2009 are negotiating the minimum of 12 Tax Information Exchange Agreements (TIEAs) required to secure upgrade to the White List. Meanwhile, however, the goal posts are moving rapidly. The OECD has already indicated that it regards 12 TIEAs as the absolute minimum and expects the bar to rise. It has also initiated a peer review process to monitor how effectively the tax havens are meeting their treaty obligations. At the same time political pressure in support of automatic information exchange has increased significantly since the 2009 G20 London Summit. By the end of 2010, the OECD intends to release its guidelines on country-by-country reporting by multinational companies, which will make it significantly easier for national tax authorities to detect profits shifting to tax havens. Cumulatively these pressures are eroding the secrecy space that many tax havens depend on to attract their clients (Hampton, 1996b).

In the course of the past decade several of the less functional SIE tax havens have retrenched from financial services (e.g. Samoa, Vanuatu and the Cook Islands: ABC Radio Australia, 2009), or closed down their OFCs entirely (Niue, Nauru and the Marshall Islands). In the Western hemisphere, the Turks and Caicos Islands were placed under direct rule from Whitehall in 2009 after a British government Commission of Enquiry concluded that there was 'a high probability of systemic corruption' by elected members of the Turks and Caicos Islands.[12] The Cayman Islands are struggling to bridge a structural budget deficit, and the British Crown Dependencies face EU pressure to replace their 'harmful' corporate tax regimes and also join the automatic information exchange system of the EU's STD. Unlike in 2001, when a friendly US administration rode to their rescue, the SIE tax havens lack powerful political allies who will intervene on their behalf.

Jersey Case Study

In 2009, in the midst of global financial market crisis, the value of deposits held in Jersey fell by approximately 20%, and the value of funds under administration fell by 30% (McRandle, 2010a). The island's OFC had already been hit the previous year by falling demand for securitised debt: two Jersey law firms, Ogiers and Mourants, were leading specialists in these debt instruments. In May 2009, a panel of economic advisers to the island's Chief Minister warned that a £60m 'black hole' structural deficit is likely by 2012 (*Jersey Evening Post*, 2009), and in October 2009 the Chief Minister reported that after consultation with a UK government minister he had been advised that 'the UK felt that other Member States are increasingly unlikely to accept their stance that the fiscal regimes in the Crown Dependencies are fully compliant with the EU Code of Conduct on Business taxation' (Le Sueur, 2009). For the first time in decades, Jersey faces the possibility that its OFC industry might not be sustainable at its current scale: a plan B is required (Body, 2010).

Jersey, however, displays many of the problems of 'collective myopia' noted by MacKinnon *et al.* (2007). Echoing Hassink's (2005, p. 253) observations elsewhere, having previously been able to rely on large taxable profits of the multinational banks and law firms, the States of Jersey were under scant pressure to attract investment in other sectors or pay anything more than lip-service to economic restructuring. The island's tourism industry has been largely crowded out (Hampton and Christensen, 2007) and talk of attracting inward investment into new sectors, including information technology, a full-service university, or 'creative industries' (McRandle, 2010b), appears wishful thinking.

At the level of functional lock-in, the OFC sector, which directly accounts for around one-quarter of the economically active population and over 50% of gross value added in the local economy, dominates the labour and commercial property market, and crowds-out prospects for diversification: 'financial capital appears to be able to out-compete other industries, particularly tourism, to gain dominance within the local political economy' (Hampton and Christensen, 2007, p. 1,014).

Considering the issue of political lock-in, Christensen and Hampton (1999, p. 186) described the sector's capture of the island's polity: 'Having established predomi-nance, the financial services sector used its political power to secure additional fiscal and regulatory advantages'. Political power is exerted through a number of players, including Jersey Finance (the OFC's marketing arm), plus a large number of industry associations representing banks, law firms, trust and company adminis-trators, hedge funds, the local branch of the Institute of Directors, the Jersey Chamber of Commerce and Industry, and powerful labour unions representing staff employed in the sector. These actors have been highly successful in securing favourable regulatory and tax treatment, even to the extent that they have been able to draft and dictate financial laws (Mitchell *et al.*, 2002). Palan *et al.* (2010, p. 187) argue that the political independence of Jersey and similar islands 'is more apparent than real, for their developmental and social goals are subject to the whim of foreign capital'. Senator Stuart Syvret, a senior politician in the States of Jersey, described its government as a 'legislature for hire' (BBC, 1996).

Concerning cognitive lock-in by decision-makers in Jersey, Hampton and Christensen (2007, p. 1,011) described how 'key players (in Jersey) failed to comprehend the long-term implications of the crowding-out issue and lacked the political power to represent their interests'. For many decades Jersey's key politicians have assumed that the EU could not extend its powers to include Crown Dependencies and that the UK government would intervene to protect their autonomy on tax matters. This is clearly no longer the case: the powers of the EU Code of Conduct Group on Business Taxation extend to the Crown Dependencies and the Group has required their respective governments to remove 'harmful tax practices' such as the 'ring-fencing' of tax-exempt status to non-resident companies.

Despite clear evidence that their existing tax regimes constituted harmful tax practices as defined by the EU, Jersey officials are largely dismissive of efforts to strengthen international cooperation. For example, the Chief Executive Office of Jersey Finance, Geoffrey Cook (2009), commented that:

> An unlikely alliance of tax hobbyists, left wing newspapers, trades unions, and development agencies has catalysed around calls for greater concentration of the means of wealth creation in the hands of governments, and implicitly greater taxation of business and wealthy individuals through the outlawing of wealth structuring and planning, together with restrictions on cross border capital flows. They hope that their own constituencies will be beneficiaries of this new 'contract', with the authors; the tax hobbyists, gaining fame and funding, and their supporters feeling validated in their enduring distrust of the wealthy and their advisers.

Despite having been warned in 2006 by a number of experts, including one of the authors of this paper,[13] that proposed amendments to their corporate tax regime (the so-called Zero-Ten tax policy) would be rejected by the Code of Conduct Group, in 2007 the States of Jersey adopted measures that were indeed deemed unacceptable in 2009.

Similarly, the Jersey authorities continue to refuse to adopt automatic information exchange within the framework of the EU's Savings Tax Directive. This decision, confirmed in November 2009, runs counter to their claims to be a transparent and well-regulated jurisdiction that cooperates with combating cross-border crime. In practice, Jersey's lack of financial market transparency, its weak international treaty arrangements for information exchange and its unwillingness to cooperate with the EU led to its being ranked eleventh out of 60 in Tax Justice Network's 2009 Financial Secrecy Index (Financial Secrecy Index, 2009). Although the tide appears to have shifted significantly against tax avoidance (Blackhurst, 2010), the Jersey authorities remain committed to business as usual, indeed during discussions in March 2010 between one author and Members of the States of Jersey, he was told that the expectation is that a change of UK government in mid-2010 will reverse the tide of EU measures against tax havens;[14] but even so, Jersey has been badly affected by the severe tax competition between the Crown Dependencies, and now faces having to implement considerable tax hikes in the short- to medium-term to cover the missing revenue.

Conclusions

Tax havens have, in the recent past, shown an ability to resist successfully external attempts to close down their activities (Palan *et al.*, 2010). However, the powerful coalitions of interests involved in initiatives launched since the mid-2000s are strongly motivated to take action, not least because of the budgetary crises facing so many developed and less-developed countries. For this reason it would be unwise for small islands hosting significant OFCs to ignore the stronger international cooperation measures being promoted by the EU and the OECD, and the potential changes to international financial reporting standards for multinational companies. Cumulatively these measures could radically strengthen the international financial architecture and significantly increase transparency, which will inevitably diminish opportunities for corporations and individuals to use tax havens for tax avoidance. It would equally be unwise for small islands to count on being able to muster the political support they were able to draw upon when resisting the 2000 OECD anti-tax haven initiative (Palan *et al.*, 2010), as the focus for political change no longer lies on blacklisting specific jurisdictions, but on targeted measures designed to increase financial market transparency and strengthen international cooperation on tax matters. Examples of such measures include the tax information exchange treaties being promoted by the G20 and the OECD, and the international financial reporting standard for country-by-country reporting by multinational companies that the OECD has committed to adopt as a guideline before the end of 2010 (Houlder, 2010b).

Some tax haven islands, including Jersey, are already facing unprecedented budgetary pressures; but they have limited scope for reducing their dependence on offshore financial services. With approximately one-quarter of its economically active population directly employed in the OFC, and the majority of the remaining workforce employed in secondary sectors such as construction, distributive trades and catering, there is virtually no alternative skills base on which new industries can draw. This path dependence has been reinforced by the extraordinarily high costs of land and labour, which have crowded-out pre-existing industries. Taking measures to diversify the local economy will therefore require politically unpalatable steps to reduce the domestic cost base significantly.

Unlike Vlcek (2008)—who appears broadly optimistic, arguing that the 'pessimism' of Hampton and Christensen (2002) concerning the OFCs' future did not happen and that they are in fact thriving—we have argued here that the present context is radically altered. Specifically, this paper has argued that tax havens hosted in small islands cannot ignore the exogenous shock to global financial capitalism caused by banking crises in 2007 and 2008, or the subsequent economic crises that have engulfed many countries in what is likely to be the most protracted recession since the 1930s. Recognition of this new world, and the implicit exogenous shock, is even seen in the (otherwise arguably timid) Foot Review (Foot, 2009a, 2009b) of the UK's offshore centres.[15]

Small island hosts of offshore finance cannot just hope that the status quo might somehow be maintained, or that the new international coalitions will somehow dissipate, leaving them free to remain as hosts of lucrative tax haven activity. Path dependency theory would suggest that the past actions and policy choices of the small island hosts themselves have contributed significantly to the serious predicament that they now find themselves in, and consequently the extremely limited economic development possibilities that remain open to them.

We would suggest that the future may be austere for many small island hosts of offshore finance. In the worst case scenario, islands could face a crash of real estate and land prices, significant emigration off-island where the most mobile leave (as in the recent cases of Iceland and Ireland) and a deep financial crisis of the local state. For non-independent island economies such as the UK Overseas Territories—or conceivably Crown Dependencies such as Jersey—this could result in increased direct control from the UK. For independent small island jurisdictions, the economic crisis could require IMF emergency loans to keep the economy afloat. In many cases, we would suggest that the outlook for small island hosts of offshore finance is bleak as there is scant evidence of the existence of a practical or realistic alternative 'plan B'.

Notes

1. A comment on the ECJ ruling is available at http://www.internationaltaxreview.com/?Page= 10&PUBID=35&ISS=21603&SID=622016&TYPE=20
2. The speech is available at http://www.hm-treasury.gov.uk/speech_fst_270110.htm
3. Approaches range from conventional neoclassical orthodoxies (World Bank, 2005; Armstrong and Read, 2006) via 'vulnerabilities' (Briguglio, 1995) to broadly political economy type writings (Hampton and Christensen, 2002; Baldacchino, 2006).
4. Offshore finance and tax havens have earlier origins with notable and organised commercial activity from at least the 1920s (Hampton, 1996a) but significant changes in the scale and scope of activities dates from the 1960s.
5. The argument over definitions of 'tax havens' versus 'offshore finance centres' has been discussed across the literature (see, for example, Hampton, 1996a; Palan et al., 2010) and by civil society groups such as TJN who have popularised the term 'secrecy jurisdictions' (see www.secrecyjurisdictions.com).
6. This description is given to themselves by the so-called leading offshore legal firms. It appears in recruitment adverts, e.g. on the www.legalweekjobs.com website. One job advert dated 22 February 2010 (US$180–210,000 'tax free') was entitled: 'Senior Structured Finance Associate—Cayman Islands—Offshore Magic Circle Client'.
7. Approximately 120 civil society organisations, including development non-governmental organisations, human rights organisations, trade unions, faith groups and others, agreed by an overwhelming majority at a meeting in Paris on 12 January 2009 that tackling tax havens was the number one priority for campaigning around the April 2009 G20 summit in London. Most recently (December 2010) tax avoidance campaigners from UK Uncut targeted Topshop outlets in major UK cities. The owners of Topshop, Sir Philip Green and his wife, were heavily criticised for their use of the Monaco tax haven to avoid UK taxes on dividends (BBC, 2010).
8. See www.financialsecrecyindex.com
9. See http://register.consilium.europa.eu/pdf/en/09/st16/st16473-re01.en09.pdf
10. In January 2010 the Tax Justice Network, working in partnership with the Inter-American Centre of Tax Administrations, launched a pilot project involving three Latin American countries that have indicated interest in negotiating multilateral information exchange treaties with the EU countries.
11. See civil society letter to G20 Finance Ministers on 28 October 2009 in the run-up to the G20 Saint Andrews conference, signed by Eurodad, LatinDad, TJN, CIDSE, GFI, Christian Aid, Oxfam, Action Aid and the Plateforme Paradis Fiscaux et Judiciaires.
12. The Commission of Enquiry report is available at http://www.fco.gov.uk/en/news/latest-news/ ?view=PressS&id=20517220
13. John Christensen was Economic Advisor to the States of Jersey from 1987 to 1998.
14. Personal communication with John Christensen, March 2010.
15. Interestingly, the Foot Review remit did not include London, arguably the largest and most significant offshore finance centre (and tax haven) in the global financial system.

References

ABC Radio Australia (2009) Tax haven clean-up hurts the Pacific, 27 August, http://www. radioaustralia.net.au/pacbeat/stories/200908/s2668920.htm, accessed 23 March 2011.

Acemoglu, D., Johnson, S. and Robinson, J. (2002) Reversal of fortune: geography and institutions in the making of the modern world income distribution, *The Quarterly Journal of Economics*, 117(4), pp. 1,231–1,294.

Armstrong, H. and Read, R. (2006) Geographical 'handicaps' and small states: some implications for the Pacific from a global perspective, *Asia Pacific Viewpoint*, 47(1), pp. 79–92.

Austin, G. (2008) The 'reversal of fortune' thesis and the compression of history: perspectives from African and comparative economic history, *Journal of International Development*, 20, pp. 996–1,027.

Baldacchino, G. (2006) Managing the hinterland beyond: two ideal-type strategies of economic development for small island territories, *Asia Pacific Viewpoint*, 47(1), pp. 45–60.

BBC (1996) Close up: a storm in a haven, broadcast on BBC2, 13 November.

BBC (2010) Topshop's flagship London store hit by tax protest, BBC News, 4 December, http://www.bbc.co.uk/news/uk-11918873

Bertram, G. (2007) Reappraising the legacy of colonialism: a response to Feyrer and Sacerdote, *Island Studies Journal*, 2(2), pp. 239–254.

Blackhurst, C. (2010) How the tide turned against the tax avoiders, *London Evening Standard*, 11 March, http://www.thisislondon.co.uk/standard/article-23814309-how-the-tide-turned-against-the-tax-avoiders.do

Body, P. (2010) What's in store now that the UK sees any kind of blatant tax planning as unacceptable? *Jersey Evening Post*, 18 February, http://www.thisisjersey.com/2010/02/18/whats-in-store-now-that-the-uk-sees-any-kind-of-blatant-tax-planning-as-unacceptable/

Briguglio, L. (1995) Small island states and their economic vulnerabilities, *World Development*, 23(9), pp. 1,615–1,632.

Christensen, J. and Hampton, M. P. (1999) A legislature for hire: the capture of the state by Jersey's offshore financial centre, in M. Hampton and J. Abbott (Eds), *Offshore Finance Centres and Tax Havens: The Rise of Global Capital* (Basingstoke: Macmillan), pp. 166–191.

Cook, G. (2009) A taxing fundamentalism. Comment by Geoff Cook, Jersey Finance website, http://www.jerseyfinance.je/Media/Comments-from-Geoff-Cook/A-Taxing-Fundamentalism/

Evans, D. (2009) Coca-Cola, Oracle, Intel Use Cayman Islands to avoid U.S. taxes, *Bloomberg*, 5 May, http://www.bloomberg.com/apps/news?pid=20601103&sid=aWoQkk2WY1oc

Feyrer, J. and Sacerdote, B. (2009) Colonialism and modern income: islands as natural experiments, *The Review of Economics and Statistics*, 91(2), pp. 245–262.

Financial Secrecy Index (2009) http://www.financial.secrecyindex.com

Financial Times (2009) Closing the havens. Editorial, 16 August.

Foot, M. (2009a) *Progress Report of the Independent Review of British Offshore Ffinancial Centres*, April (London: HM Treasury).

Foot, M. (2009b) *Final Report of the Independent Review of British Offshore Financial Centres*, October (London: HM Treasury).

Global Financial Integrity, Christian Aid, Global Witness, Tax Justice Network (2009) The links between tax evasion and corruption: how the G20 should tackle illicit financial flows, Civil Society proposals to the G-20 in advance of the Pittsburgh Summit, http://www.financialtaskforce.org/wp-content/uploads/2009/09/illicit_financial_flows_asks_for_g20-2.pdf

Grabher, G. (1993) The weakness of strong ties: the lock-in of regional development in the Ruhr area, in G. Grabher (Ed.), *The Embedded Firm: On the Socio-economics of Industrial Networks* (London: Routledge).

Gurria, A. (2009) The end of the tax haven era, *The Guardian*, 31 August, http://www.guardian.co.uk/commentisfree/2009/aug/31/economic-crisis-tax-evasion

Hampton, M. P. (1996a) *The Offshore Interface. Tax Havens in the Global Economy* (Basingstoke: Macmillan).

Hampton, M. P. (1996b) Creating spaces. The political economy of island offshore finance centres: the case of Jersey, *Geographische Zeitschrift*, 84(2), pp. 103–113.

Hampton, M. P. and Christensen, J. (2002) Offshore pariahs? Small island economies, tax havens and the re-configuration of global finance, *World Development*, 30(9), pp. 1,657–1,673.

Hampton, M. P. and Christensen, J. (2007) Competing industries in islands: a new tourism approach, *Annals of Tourism Research*, 34(4), pp. 998–1,020.

Hassink, R. (2005) How to unlock regional economies from path dependency? From learning region to learning cluster, *European Planning Studies*, 13(4), pp. 521–535.

Hay, R. J. (2005) Beyond a level playing field: free(r) trade in financial services. Paper presented at the *STEP Symposium*, London, 19–20 September, p. 6.

Houlder, V. (2009) Ports in a storm, *Financial Times*, 18 November.

Houlder, V. (2010a) OECD hails tax haven crackdown, *Financial Times*, 19 January.

Houlder, V. (2010b) Treasury aims to end tax outflows, *Financial Times*, 11 February, http://www.ft.com/cms/s/0/e070175a-163c-11df-8d0f-00144feab49a.html

Hutchinson-Jafar, L. (2009) Caribbean fears witch-hunt in tax haven crackdown, *Jamaica Observer*, 8 May, http://www.jamaicaobserver.com/magazines/Business/html/20090507T220000-0500_151005_OBS_CARIBBEAN_FEARS_WITCH_HUNT_IN_TAX_HAVEN_CRACKDOWN.asp

International Herald Tribune (2009) Editorial: 'If Switzerland can ...', 21 August, p. A16.

Jersey Evening Post (2009) Another black hole, 6 May, www.thisisjersey.com/2009/05/06/another-black-hole/

La Tribune (2009) Ils veulent en finir avec les paradis fiscaux, 4 March, p. 1.

Le Monde (2009) Londres ou New York sont aussi des paradis fiscaux, interview with John Christensen, 24 March, http://www.lemonde.fr/la-crise-financiere/article/2009/03/24/john-christensen-londres-ou-new-york-sont-aussi-des-paradis-fiscaux_1171954_1101386.html

Le Sueur, T. (2009) Letter to members of the States of Jersey relating to discussions with UK government about the position of the EU Code of Conduct Group on Business Taxation, www.taxresearch.org.uk/Blog/2009/10/14/the-crown-dependencies-do-not-comply-with-the-eu-code-of-conduct/

MacKinnon, D., Cumbers, A., Pike, A. and Birch, K. (2007) *Evolution in Economic Geography: Institutions, Regional Adaptation and Political Economy*, Working Paper 12, November, Centre for Public Policy for Regions, University of Glasgow.

Markoff, A. (2009) Part 1: the early years—1960's: the Cayman Islands: from obscurity to offshore giant, *Cayman Financial Review*, 14 (first quarter), http://www.compasscayman.com/cfr/cfr.aspx?id=108

Martin, R. and Sunley, P. (2006) Path dependence and the evolution of the economic landscape, *Journal of Economic Geography*, 6(4), pp. 395–438.

Mathiason, N. (2009) Financial hurricane shakes the tax havens, *The Observer*, Business Section, 6 September, p. 7.

McRandle, H. (2010a) Fall in deposits and funds last year, *Jersey Evening Post*, 26 February, http://www.thisisjersey.com/2010/02/26/sharp-fall-in-deposits-and-funds-last-year/

McRandle, H. (2010b) Creative industries to replace finance? *Jersey Evening Post*, 11 March, www.thisisjersey.com/2010/03/11/creative-industries-to-replace-finance/, accessed 14 March 2010.

Mitchell, A., Sikka, P., Christensen, J., Morris, P. and Filling, S. (2002) *No Accounting for Tax Havens* (Basildon, Essex: Association for Accountancy and Business Affairs).

Mitchell, D. (2001) CFP strategic memo on OECD strategy, Centre for Freedom and Prosperity, Washington, 15 May, http://www.freedomandprosperity.org/Papers/m05-15-01/m05-15-01.shtml, accessed 14 March 2010.

O'Brien, R. (1992) *Global Financial Integration. The End of Geography* (London: Pinter).

OECD (1998) *Harmful Tax Competition: An Emerging Global Issue* (Paris: OECD).

Palan, R. (2003) *The Offshore World. Sovereign Markets, Virtual Places and Nomad Millionaires* (Ithaca: Cornell University Press).

Palan, R., Murphy, R. and Chavagneux, C. (2010) *Tax Havens: How Globalization Really Works* (Ithaca: Cornell University Press).

Park, Y. (1982) The economics of offshore finance centres, *Columbia Journal of World Business*, 17(4), pp. 31–35.

Quérée, B. (2009) Tax crisis as EU attacks, *Jersey Evening Post*, 14 October, p. 1.

Sanders, R. (2002) The fight against fiscal colonialism. The OECD and small jurisdictions, *The Round Table*, 91(365), pp. 325–348.

Scott, P. (2001) Path dependence and Britain's 'coal wagon problem', *Explorations in Economic History*, 38(3), pp. 366–385.

Sharman, J. (2006) *Havens in a Storm. The Struggle for Global Tax Regulation* (Ithaca, NY: Cornell University Press).

Tax Justice Network (2005) *The Price of Offshore* (London: TJN).

Vlcek, W. (2007) Why worry? The impact of the OECD harmful tax competition initiative on Caribbean offshore financial centres, *The Round Table*, 96(390), pp. 331–346.

Vlcek, W. (2008) Competitive or coercive? The experience of Caribbean offshore financial centres with global governance, *The Round Table*, 97(396), pp. 439–452.

World Bank (2005) *A Time to Choose. Caribbean Development in the 21st Century*, Report No. 31725-LAC (Washington, DC: World Bank).

Zaki, M. (2010) *Le Secret Bancaire est Mort, Vive l'évasion Fiscale?* (Lausanne: Éditions Favre).

Between a Rock and a Hard Place: Small States in the EU–SADC EPA Negotiations

BRENDAN VICKERS

Head of Research and Policy, International and Economic Development, Department of Trade and Industry, South Africa

ABSTRACT *This article explores the role and effectiveness of small state trade diplomacy in the negotiations to conclude Economic Partnership Agreements (EPAs) between the European Union (EU) and the African, Caribbean and Pacific (ACP) group of countries, focusing specifically on the Southern African Development Community (SADC). Given the vast power asymmetries between the EU and the ACP, small states have had limited bargaining power to shape the process and the outcome of the negotiations. Unlike most other ACP EPA negotiations, the SADC small states were also caught between a rock (EU) and a hard place (South Africa), with both parties competing to promote their visions for regional integration. In the end, the EPA process split SADC into four sets of separate trade regimes with the EU, undermining the established regional integration project. The article explains this divisive outcome of the SADC EPA process by analysing the negotiation behaviour of the main parties, specifically the 'weaker' players. The article concludes with key lessons for small states' future trade negotiations.*

Introduction

Small vulnerable economies (SVEs) are today emerging as novel players in the international political economy (see Narlikar, this issue). Not only have they adopted a range of creative strategies to ameliorate their structural vulnerability at the margins of the world economy, but they are also contesting global governance more broadly (Cooper and Shaw, 2009). Their diplomatic resilience in the face of 'smallness' is particularly promising in the World Trade Organisation (WTO). In this case, coalition-formation by smaller developing states has transformed their participant status from passive 'objects' of trade talks to 'subjects' in the WTO's Doha Development Agenda, notably in the cotton negotiations (Lee and Smith, 2008). Compared with this proactive agency in multilateral trade, however, small state diplomacy has been less effective in bilateral bargaining, where vast power asymmetries constrain the possibilities for more assertive behaviour. Within this context of economic asymmetries, William Zartman and Jeffrey Rubin (2002, p. 16) predicted that: 'Negotiators with high relative power tend to behave manipulatively and exploitatively, whereas those with perceived lower power tend to behave submissively'. This proposition is broadly supported by small developing countries' experience with bilateral trade negotiations, where power asymmetries between the strong and the weak mean that small states are mostly 'acted upon' by the bigger parties (see Oxfam, 2007; UNCTAD, 2007). As the more powerful parties are able to control the negotiation process and obtain results to their liking, the negotiating endgame tends to 'confirm a given power distribution' (Zartman and Rubin, 2002, p. 4).

The experience of the Economic Partnership Agreement (EPA) negotiations conducted between the European Union (EU) and 79 former colonies from the African, Caribbean and Pacific (ACP) region—including numerous Commonwealth member countries—illustrates this quandary. Confronted by the EU's vast 'shadow of power', the EPA negotiations have been a bruising battle for the ACP. By the WTO's deadline of 31 December 2007, only the CARIFORUM countries (CARICOM states plus the Dominican Republic) had finalised a comprehensive EPA covering not only trade in goods, but also the liberalisation of services and investment, competition, intellectual property rights, e-commerce and trade-related measures (Mohammed, 2009). Elsewhere, fewer than half the ACP countries had initialled 'goods only' interim agreements to avoid trade disruptions. Those ACP countries remaining outside the EPA, without any agreement with Brussels, were shut out of the European market through stricter trade disciplines (i.e. higher import duties or more restrictive rules of origin). Since the deadline, several other ACP countries have initialled or signed full or interim agreements.

With their small markets, narrow capital and production bases, and undiversified economies locked into commodity-dependent export paths (e.g. bananas, beef, horticulture, sugar and tuna), the ACP countries wield considerably less bargaining power than Europe. The Pacific Islands, the smallest group in the EPA negotiations, has a combined gross domestic product (GDP) of only US$9bn, which is 1,400 times smaller than the EU's. Even the largest group, the West Africa region, is 80 times smaller than the EU in terms of GDP (Oxfam, 2006, p. 2). Moreover, with nearly half of ACP exports destined for Europe, far more is at stake for smaller states than for the giant European economy. Clearly, small states' economic vulnerabilities, material inequalities and trade dependencies *vis-à-vis* larger parties render them weaker actors in bilateral settings. By contrast, the EU has held most of the cards: market power (i.e. access to the vast common market, with a combined GDP of US$13,300bn); financial power (i.e. development assistance through the European Development Fund and other aid facilities); and negotiating power, with considerable diplomatic resources and experience from having negotiated a series of free trade agreements (FTAs). These asymmetries between the EU and ACP have had an impact on the final negotiated structure and content of the EPAs, which have broadly favoured Brussels' interests (although Southern African states have won some important concessions, as is later discussed). Understandably, many ACP countries have been perturbed by the EPA outcome, some even calling for new negotiations. This reflects an appreciation that these new trade agreements can potentially have a large *direct* impact on their developmental prospects, whether positive or negative, and their resilience in the face of the multiple economic, finance, food and climatic crises of the past decade (Jones and Marti, 2009).

Of all the ACP regions, the Southern African EPA process has been the most acrimonious and divisive. With South Africa's late inclusion as an EPA party, the region's smaller states confronted two relative 'shadows of power'—the EU and South Africa—each competing to promote their own visions for regional integration. In the end, the EPA process split the 15 members of the Southern African Development Community (SADC) into four sets of separate trade regimes with the EU, each varying considerably from one another. This dispiriting scenario has raised concerns about the future of SADC's regional integration project. This article seeks to understand and explain these regional divisions, and the role and effectiveness of small state trade diplomacy in the face of two 'shadows of power'. The article proceeds in three steps. The next section provides a brief overview of regional integration dynamics in SADC, in order to appreciate the significance of the EPA divisions that subsequently emerged. The following section seeks to explain the regional disarray in Southern Africa by analysing the negotiating behaviour of the main SADC EPA protagonists. The article concludes by drawing lessons from the SADC EPA experience for small states—and international trade negotiations more broadly.

The Southern African Region: One Gulliver, Many Lilliputians

Established in 1992, SADC is a formal Regional Economic Community (REC) comprising 15 developing countries, largely centred on South Africa as the regional economic powerhouse. Its membership varies considerably according to levels of economic development and population and geographic size (ranging from large land states to the small island developing states of Mauritius and Seychelles). More than half of the SADC economies rank as least-developed countries (LDCs), while the remainder are small, vulnerable and still dependent on natural resources and commodity exports to fuel growth. South Africa is the *de facto*, if reluctant, hegemon: the country accounts for about 60% of all intra-SADC trade and about 70% of its GDP, while South African firms dominate the region's services sectors (Grobbelaar and Besada, 2007). South Africa is the continent's leading foreign direct investor, more invested even than China (UNCTAD, 2010).

Being small and marginal players in the global economy, Southern African countries have prioritised deeper regional integration as a means of stimulating growth and reducing poverty. Today, a conventional market integration paradigm underpins SADC's regional integration agenda: an FTA by 2008, a customs union by 2010, a common market by 2015, a monetary union by 2016 and a single currency by 2018. Notwithstanding the establishment of the FTA, intra-regional trade remains low. The region's stagnant trade pattern is explained by three challenges: an underdeveloped and non-diversified industrial manufacturing base in most SADC countries, inadequate infrastructure to support trade in goods, and non-tariff barriers (NTBs) that inhibit market access. There has also been some progress towards liberalising trade in services, with the stated ambition of building regional convergence in rules before opening up to external parties, which the EC has demanded in the EPA negotiations (discussed later). Within SADC, the nucleus of regional integration is the Southern African Customs Union (SACU), comprising South Africa and its lesser developed neighbours Botswana, Lesotho, Namibia and Swaziland (BLNS). Given South Africa's regional economic dominance, the latter are 'locked' into importing relatively high-cost services from South Africa, given that most key South African service sector markets display concentrated oligopolistic market structures. The BLNS are thus seeking to diversify their trade and investment relations, and introduce external competition into their markets, including from the EU (Draper *et al.*, 2007).

At the outset of the SADC EPA negotiations, two issues were thus pertinent. The first was how to design an appropriate 'trade for development' partnership that safeguards Southern Africa's preferential margins in the EU market, together with aid-for-trade to address regional production, supply-side and trade governance constraints. Notwithstanding the overtures from China and India, the EU remains the region's single largest trade, investment and development aid partner. Under the Lomé/Cotonou Conventions (I–IV, 1975–2000), SADC countries—excluding South Africa—have historically enjoyed preferential access to the EU market over their competitors. With the exception of the regional hegemon, Southern African exports to Europe are concentrated in a few undifferentiated commodities, including diamonds, beef, fruits, fish, sugar and petroleum. Given South Africa's relative competitiveness, in 1999 Brussels and Pretoria concluded a separate Trade, Development and Cooperation Agreement (TDCA), which will be fully operative by 2012. Although the BLNS countries were not part of these negotiations, the TDCA has a clear impact on them by virtue of SACU's common external tariff, effectively making them *de facto* members. One of South Africa's motivations for joining the EPA negotiations in 2004 was thus to consolidate the region's trading relations internally and *vis-à-vis* the EU.

The second point is more instructive, namely that SADC has adopted clear regional integration benchmarks, flanked by both hortatory (SADC) and legal (SACU) obligations to negotiate collectively with external trading partners. However, from the outset of the SADC EPA process, the domestic interests of SADC parties—*inter alia* maintaining preferential access to the EU market—trumped regional coherence and collective representation. The resulting intra-regional divergence thus generated considerable doubts over the future of SACU as the world's oldest functioning customs union and even an SADC-wide customs union (see Draper and Khumalo, 2009). Although many of these concerns have now been resolved and an SADC consolidation agenda on the cards, the next section explains the initial EPA outcome by analysing the negotiation behaviour of the parties.

Exploring the Negotiation Behaviour of the SADC EPA Parties

As Amrita Narlikar (2010) has argued, the negotiation behaviour of any state can, for analytic purposes, be usefully divided into several variables, *inter alia* normative framing, coalition-building, negotiation strategy and leadership. Using the first three of these variables, the following section explores the bargaining behaviour of the main protagonists.

Normative Framing

The ideological or normative frames of the various negotiating parties played a key role in shaping the outcome of the SADC EPA. Paradoxically, EC and SADC trade diplomats both framed their negotiating positions in terms of the Cotonou Agreement's objectives to integrate gradually ACP countries into the world economy, promote sustainable development and reduce poverty. Notwithstanding this common point of reference, Europe (represented by the EC) and its former colonies have been at loggerheads over how best to design a 'trade for development' partnership that encompasses all of Cotonou's objectives. The EC's negotiating position was based on a neoclassical 'problem-solving' perspective, equating freer trade, investment liberalisation and competition with development. By contrast, the ACP countries adopted a more interrogatory posture, placing greater emphasis on the Cotonou Agreement's transformational objectives, rather than simply trade *per se*.

In the final analysis, the structure, content and framing of the EPAs have been heavily weighted in favour of the EU's interests. From Brussels' perspective, the developmental concerns of the ACP countries can be accommodated within the framework of reciprocal FTAs, with special and differential treatment (SDT) at the margins. Led by former European Trade Commissioner Peter Mandelson, DG Trade hence adopted a rather literal (i.e. textual) approach to the interpretation of WTO law, notably the reciprocity requirements of GATT Article XXIV (Ochieng, 2007). The need for reciprocity in the EPAs was motivated by the apparent 'failure' of three decades of preferential aid and trade under the EU–ACP Lomé/Cotonou Conventions. Prior to the release of its 2020 Trade Strategy in November 2010 (European Commission, 2010), the grand centrepiece of the EC's trade agenda was 'Global Europe' (European Commission, 2006). As the world's leading exporter of services, with 28% (European Services Forum (ESF), 2007), the EU has sought to position itself as the leader of global services trade. 'Global Europe' regards the export of regulation as one way of maximising the competitiveness and market share of European services and other firms in the world economy. Notwithstanding widespread ACP opposition to WTO-plus rule-making, EC trade negotiators hence insisted that the EPAs must be comprehensive in their coverage, including the so-called 'Singapore Issues' (i.e. services, investment, competition, government procurement and trade facilitation).

By contrast, the ACP countries have adopted a more teological (i.e. holistic) approach to the interpretation of WTO law, as the latter outlines numerous SDT and other provisions for poorer countries (Ochieng, 2007). ACP trade diplomats thus appealed to the developmental norms embedded in the global trading regime as the appropriate framing for the negotiations. Brussels indeed stood accused of Janus-faced politics: in bilateral negotiations where power asymmetries are amplified, the EC has demanded reciprocity from the world's poorest nations, whereas in the WTO the EC has proposed a 'round for free' for LDCs (Elsig, 2009). Many ACP countries, including SADC, argued that there was no basis for reciprocity in North–South RTAs, as GATT Article XXIV was a post-war arrangement tailored to developed country interests. They furthermore insisted that there was no obligation to negotiate the Singapore Issues to ensure a WTO-compliant trade agreement with Europe. Indeed, the bulk of the new generation issues have been suspended under the WTO's Doha Round, with the exception of trade facilitation. In a signature soft power strategy, developing countries have hence sought—unsuccessfully—to shift the debate and focus of the EPA negotiations from their narrow mercantilist preoccupations to a 'trade for development' partnership.

In the case of Southern Africa, the latter has found fruition in the notion of 'developmental regionalism', which is increasingly in vogue as an alternative to the EU's market integration paradigm. Rather than simply focusing on tariffs as the greatest barriers to intra-regional trade flows in SADC, this model of regionalism seeks to address regional production, supply-side and trade governance constraints as a precursor to deeper market integration and socio-economic development. However, SADC's hopes of harnessing a partnership with the EU and the development and assistance facilities in the EPA to resolve these core challenges and to strengthen regional integration were soon dashed by the EPA's divisive politics. As the subsequent sections argue, maintaining countries' market access preferences into the EU ultimately came at a high price for regional coherence in Southern Africa.

Regional Coalition-building

The second factor to consider is the role and contribution of regional coalition-building in the SADC EPA negotiations. By virtue of their weaker bargaining power, collective action and representation by ACP countries could lead to more leverage *vis-à-vis* the larger party, Europe. Theoretically, alignment with a more developed regional partner (e.g. Ghana, Kenya, Nigeria or South Africa) may even strengthen collective propositions; but the latter may also lead to greater divergence and mistrust by introducing a second frontier of contestation in the negotiating equation, particularly where the dominant partners pursue their own interests. Compared with the more instrumental episodes of issue-based diplomacy in the WTO and elsewhere, where small states have been particularly effective, regional bloc-type coalitions have performed more poorly (Narlikar, 2003). The EPA negotiations were none the less conducted on a regional basis, with the EU playing 'hub' to six ACP regional 'spokes', four of which are in Africa and the other two in the Caribbean and Pacific. From the outset of the SADC EPA negotiations, the process was dominated by intra-regional divergence and mistrust, rather than the rational convergence of domestic trade policy positions. Forging an Africa-wide continental consensus as a counterweight to Europe was even more daunting. The African Union's High-level Panel (2007, p. 76) Audit Report lamented this fact:

In 2006, Member States and the recognised Regional Economic Communities abandoned a common African position on the Economic Partnership Agreements (EPAs). Despite widespread analysis and public perception that the EPAs offer less than the previous Cotonou Agreement and are potentially disastrous for Africa's fledgling industries, domestic public revenue base and agriculture, Africa has faced the European Commission with contradictory and divided configurations.

Instead of engaging Brussels' trade diplomats as a consolidated REC, SADC's 15 member states initially splintered into two separate configurations—and later four (see Table 1). Confusingly, the resulting EPA groupings were not coterminous with the boundaries of the existing SADC REC, which the African Union formally recognises as a building-block of the African Economic Community. Although Article 37(5) of the Cotonou Agreement enjoined the ACP countries to self-designate their own negotiating groups, the resulting divisions among SADC countries undermined the region's integration ambitions. In so doing, the SADC EPA process exposed the underlying weaknesses of the regional integration agenda in Southern Africa, already evident from the 'spaghetti bowl' phenomenon of overlapping REC memberships and divergent loyalties (see Bhagwati, 2008). The EC was also partly complicit in this regional balkanisation. Through subtle directives or pressures on parties to join particular negotiating configurations, or by virtue of the Commission's substantial leverage as a leading donor in the region, the EC had a hand in reshaping Southern African geography.

The bulk of Southern African countries, particularly the LDCs and lesser-developed countries, thus chose to join the Eastern and Southern African (ESA) EPA centred on SADC's 'competitor', the Common Market for Eastern and Southern Africa (COMESA). Compared with the 'weaker' SADC Secretariat, which at the time was undergoing restructuring, the more proactive role of the COMESA Secretariat in organising and facilitating the ESA process provided a natural gravitational pull for anxious smaller countries. In addition, several countries were uneasy about South Africa's potential dominance and role in the negotiations and the prospect of benchmarking the SADC EPA against the latter's own Trade, Development and Cooperation Agreement with the EU (Shilimela, 2008). The remaining Southern African countries constituted the 'SADC-minus' group, consisting of the BLNS countries plus Mozambique, Angola and Tanzania. South Africa initially participated in this configuration as an observer only, later becoming an active negotiating party. Tanzania subsequently joined the East African Community (EAC) EPA negotiations.

Table 1. The SADC REC and the EPA negotiating configurations

SADC REC member states	SADC-minus	ESA	EAC	Central Africa
Angola	X			
Botswana	X			
Democratic Republic of Congo				X
Lesotho	X			
Madagascar		X		
Malawi		X		
Mauritius		X		
Mozambique	X			
Namibia	X			
Seychelles		X		
South Africa	X			
Swaziland				
Tanzania			X	
Zambia		X		
Zimbabwe		X		

Note: Author's own compilation.

Negotiating Strategy

The final variable to consider is the type of negotiating strategy adopted by EC and SADC trade diplomats, and how this has combined with normative framing to shape the acrimonious outcome of the SADC EPA negotiations. Negotiation strategies vary across a spectrum, ranging from distributive (i.e. value-claiming) to integrative (i.e. value-creating). Distributive strategies include tactics such as refusing to make any concessions, threatening to hold others' issues hostage, issuing threats and penalties or worsening the other party's best alternative to negotiated agreement (BATNA). Integrative strategies comprise attempts to widen the issue space and explore common solutions, i.e. strategies designed to expand rather than split the pie (Odell, 2000).

Distributive bargaining by both parties, coupled with the pressures of a pending WTO deadline, were the bane of much of the SADC EPA negotiations. Following the formalisation of the SADC EPA group, trade ministers conducted 18 months of regional consultations, between 2004 and 2006, to develop a common negotiating framework. The latter was submitted to the EC on 7 March 2006, setting out a combination of offensive and defensive demands. Most importantly, South Africa would be included as a full negotiating party alongside the smaller states in order to harmonise the region's trading relations with the EU, and also to draw on Pretoria's experience of negotiating the TDCA with Brussels. Of course, South Africa's decision to join the SADC EPA was not simply motivated by altruistic intentions. The country is intent on improving its market access into Europe, especially for agricultural commodities, which attract far higher applied duties than non-agricultural products. In one estimate, 32% of agricultural imports carry duty rates higher than 10% (Pant, 2009), providing a major incentive for South Africa to seek more favourable terms.

Second, SADC insisted that the TDCA, which *de facto* also applied to the BLNS countries by virtue of SACU's common external tariff, settled the EC's demand for reciprocity under WTO law. As the BLNS had not been party to the TDCA negotiations, the EPA would have to be sufficiently flexible to protect their sensitive domestic sectors, as well as accommodate Lesotho's LDC status. Given that the remaining SADC EPA parties were all LDCs (namely Mozambique, Angola and Tanzania), they would be exempted from reciprocity and entitled to the everything-but-arms (EBA) benefits of duty-free quota-free (DFQF) market access into the EU. In order to consolidate the region's trading relations internally and *vis-à-vis* the EU (i.e. EPA, EBA and TDCA), SADC proposed DFQF market access for all its members, including South Africa. Finally, SADC countries would seek a cooperative partnership with the EC to build national capacity on the new generation trade issues, without any obligation to negotiate the EC's regulatory agenda.

The EC procrastinated, formally responding a year later in March 2007. Brussels rejected outright the SADC EPA framework, conceding only that South Africa should be included in the SADC EPA, subject to stringent conditions. Most importantly, Brussels (backed by agricultural lobbies in Europe) insisted on nego-tiating separate market access offers for South Africa and the lesser-developed SADC states. As there is little economic basis for this differential treatment, particularly given South Africa's supply capacity constraints (Pant, 2009), the EC's defensive posturing served only to deepen existing trade policy divisions in the region.

Contrary to SADC's framing of the EPA, the EC insisted that LDCs must offer reciprocal market openings and negotiate the new generation issues to ensure WTO compliance. These demands were clearly aimed at value-claiming from smaller economies. As an extension of this distributive strategy, the EC audaciously orchestrated African business support for its negotiating agenda. This included the EC's role in establishing the Business Trade Forum EU–Southern Africa and co-drafting with the European employers' federation, BusinessEurope, a pro-EPA position for the EU–Africa Business Forum. In doing so, legitimate African business interests were completely bypassed by this political joint venture between the EC and European big business (Corporate Europe Observatory, 2009).

During the nine-month negotiating endgame between March and December 2007 (including the EC's late inclusion of contentious new issues, presented as 'deal-breakers'), positions among the SADC EPA members shifted dramatically to favour the EC's negotiating position and framing. Under the pressure of the waiver deadline, the LDCs agreed to offer the EU reciprocity; and apart from Angola, Namibia and South Africa, SADC EPA members agreed to negotiate services and investment. By the New Year, five of the SADC EPA states, namely Botswana, Lesotho, Mozambique, Namibia (with scheduled reservations) and Swaziland, initialled a goods-only interim EPA (IEPA). South Africa did not initial the agreement and continues to trade with the EU under the TDCA, while Angola's EBA preferences continue uninterrupted. South Africa and Namibia also opted out of the second phase of the negotiations covering the EC's preferred regulatory agenda (Article 67 of the IEPA). Botswana, Lesotho and Swaziland subsequently signed the IEPA on 4 June 2009, while Mozambique signed later, on 15 June 2009. Angola, Namibia and South Africa (ANSA) are thus the remaining parties outside the agreement.

For many small ACP countries, including SADC, the IEPAs represented a Faustian bargain: they were not regarded as opportunities, but the 'price' to pay to continue exporting to Europe (Bilal, 2009). Confronted by the reality of signing the IEPA or forfeiting preferential margins (and the threat of losing aid), the pressure to acquiesce to Brussels' preferred trade paradigm was palpable. Compared with former EC Trade Commissioner Peter Mandelson's confrontational bargaining and 'deal-breaker' politics, the appointment of Catherine Ashton as his successor in 2008 (and since 2010 Karel de Gught) has tempered the tone of Europe's value-claiming, heralding more integrative impulses. This change of tone also reflected internal pressures and face-saving by the Commission. A case in point was the Swakopmund round of negotiations in March 2009, where the EC's trade diplomats offered several concessions in a bid to resolve the ANSA group's outstanding concerns about the IEPA (e.g. export taxes, infant industry protection and food security). These concessions signalled a more accommodating turn in the negotiations. Notwith-standing Brussels' symbolic overtures, Namibia still withheld the country's signature because of the EC's refusal to amend the IEPA text or to provide adequate legally binding assurances on the concessions outlined in the Swakopmund Declaration. While pressure is mounting on Namibia to sign the IEPA or revert to less favourable tariff treatment for its exports, South Africa remains fully outside the agreement, largely on principled but also pragmatic grounds.

Although South Africa had raised several concerns about the IEPA (see Carim, 2009), two were particularly contentious for Pretoria. The first revolved around the

EC's insistence on the inclusion of a most-favoured nation (MFN) clause in the agreement. This represents distributive bargaining *par excellence*, as it guarantees European producers the best available access to the Southern African market in perpetuity. The MFN clause obliges SADC EPA members to grant to the EU any trade advantage offered to other major trading partners—including Brazil (Mercosur), India or China—in future trade negotiations. South Africa has argued that the MFN clause limits the ability of ACP countries to diversify their trading relations away from Europe, *inter alia* through South–South diplomacy. The South African government in particular is intent on strengthening trading relations with advanced developing countries. Reflecting this ambition, in 2009 China for the first time emerged as South Africa's biggest trading partner, overtaking its established Western markets (*Business Day*, 28 September 2009, p. 1).

Brussels' position on the MFN provision placed South Africa in a quandary. The EC insisted that MFN be linked to the DFQF market access offer it had extended to all SADC EPA states, with the exception of South Africa. The DFQF arrangement (with a phase-in period for rice and sugar, affecting mainly Mozambique and Swaziland) not only safeguards non-LDCs' preferential access into the EU market, but also greatly expands their economic opportunities by abolishing pervious duties and quotas (e.g. beef for Botswana and Namibia). At first glance, the MFN clause thus appears to be South Africa's 'price' to pay if it is committed to supporting its smaller neighbours' development; but South Africa also has strong offensive interests in advanced developing country markets. The reciprocity linchpin of trade negotiations may require greater trade-offs in favour of the country's Southern partners, rather than extending the same treatment to the EU.

As an extension of the EC's distributive bargaining, the DFQF provision also has strong mercantilist overtones. Although free access will not lead to much adjustment in Europe, as the ACP countries account for less than 3% of EU world imports and have limited supply response, the EC has secured new openings of up to 80% in ACP markets. Compared with Brussels' more lenient approach in the WTO, where it has supported a 'round for free' with minimal commitments by LDCs and SVEs, the EPAs are strongly value-claiming. This view is expressed succinctly by South Africa's chief trade negotiator: 'It appears the EC aims to ensure that no other WTO member obtains improved market access in the ACP through the multilateral process, while it secures for itself vastly improved access into the ACP economies *via* the EPA' (Carim, 2009, p. 56).

The second concern broadly revolved around the impact of the IEPA on SADC's regional integration agenda, although these concerns have now largely been resolved, with greater emphasis on consolidating both SADC and SACU. Notwithstanding the IEPA's objective to complement and support regional integration, the net effect was to fragment the SADC membership across four separate configurations. Each group negotiated market access arrangements for EU goods that varied considerably from one another, raising the spectre of trade deflection.[1] Although these market access issues have been largely resolved, the EPA divisions pose the greatest risk for developing regional regulatory regimes:

> This will complicate—and could even foreclose—efforts to foster regional integration. The separate arrangements also create the basis for new trade policy divisions in the region as they provide market opening obligations and

commitments to the EU before the region has had time to build its own regional markets and rules in such new areas as services, investment, competition and procurement. (Carim, 2009)

Moreover, the proliferation of EPA groups in the Southern African region made the establishment of an SADC-wide customs union by 2010 virtually impossible, even though that ambitious benchmark is being reviewed. The SADC IEPA had initially also raised concerns over the future of SACU a year before its centenary in 2010, with three of its five member states—Botswana, Lesotho and Swaziland—being party to a separate trade agreement with the EU, allegedly in violation of Article 31 of the 2002 SACU Agreement (which calls for collective decision-making). As a result of the differences in tariff regimes and the rules of origin under the TDCA and IEPA, South Africa had intimated that it would restrict imports covered by the agreement or raise border controls to avoid transhipment of cheap EU exports (e.g. clothing and textiles). More significantly, since the IEPA had violated SACU's common external tariff, the latter's standing under WTO rules governing customs unions had become questionable. Given the extreme fiscal dependence of the smaller SACU member states on revenue transfers from the customs union (up to 70% of Lesotho and Swaziland's annual budgets), the collapse of SACU would have spelt disaster. SADC's small states were clearly caught between a rock (i.e. loss of market access to the EU) and a hard place (i.e. loss of fiscal transfers should SACU collapse).

Recently, with the easing of pressure by the EC on the smaller SADC states to implement the IEPA, there has been a return to greater regional coherence in the negotiations. The threat to SACU has also dissipated. On several occasions, SACU Heads of State have recommitted themselves to strengthening SACU and promoting unified engagement on trade negotiations.

Table 2 summarises the negotiating behaviour of the parties. The final section seeks to draw lessons from the SADC EPA experience for small states—and international trade negotiations more broadly.

Table 2. Summary of EC and SADC EPA initial negotiating positions

	EC	SADC-minus
Normative framing	• Cotonou Agreement • 'Global Europe' strategy: new generation/regulatory agenda • Reciprocal FTAs • Literal (textual) interpretation of WTO law	• Cotonou Agreement • Developmental regionalism: production, supply-side, governance • Asymmetrical FTAs • Teological (holistic) interpretation of WTO law
Negotiating strategy	Distributive: LDC reciprocity; MFN; WTO-plus rule-making (i.e. Singapore Issues)	Distributive: asymmetry; DFQF market access; aid-for-trade
Regional coalition-building		SADC-minus SACU ANSA

Note: Author's own compilation.

Conclusion

Unlike most other ACP EPA negotiations, SADC's small states have been caught between a rock (EU) and a hard place (South Africa). The acrimony of the SADC EPA process highlighted the weak foundations of the region's integration agenda, the widely disparate nature of the region's economies, and long-simmering regional tensions and mistrust, partly related to perceptions of South Africa's regional hegemony. The result was the balkanisation of the region and serious questions over the future of SACU and even SADC, although these have been largely resolved.

More recently, the parties have narrowed their differences through more integrative bargaining, edging closer to a deal. Notwithstanding power asymmetries, the SADC states have won important concessions from the EC to support their economic development (e.g. export taxes, infant industry protection and food security). In other words, SADC states have not simply been 'submissive players' (Zartman and Rubin, 2002). Notwithstanding previous manipulative negotiating tactics, the EC has also demonstrated greater flexibility on several of the new generation issues—particularly competition and government procurement—while still insisting on binding obligations in the other areas of its comparative advantage: investment, services and intellectual property (European Commission, 2010). Interestingly, with the EC no longer pressurising the smaller signatory states to implement the IEPA, the SADC negotiating fold has demonstrated greater regional coherence and unity. Why did this whole situation arise and what can we learn from this episode of bilateral bargaining?

The first explanation concerns the apparent primacy of SADC countries' domestic interests over regional coherence and collective representation. Clearly, the decision of SADC's non-LDCs, namely Botswana, Namibia and Swaziland, to initiate the IEPA was motivated by narrow commercial interests to avoid trade disruptions. Europe's 'shadow of power' was palpable, as the BATNA for non-signatories was the stricter Generalised System of Preferences (GSP) or even higher duties, rendering their exports less competitive. Botswana, for example, risked forfeiting its preferential access for beef exports into the European market. The loss of the EU beef market (with little prospect of immediately developing alternative export channels) would have adversely affected Botswana's cattle-rearing community, especially the rural poor. By one estimate, 62% of the country's rural population depend on cattle farming; moreover, about 80% of the total cattle population is produced by small farmers, whereas only 20% is produced by commercial farmers (*Business Day*, 17 July 2009, p. 3). In contrast to South Africa's opposition to services openings, Botswana trade diplomats justified their decision to negotiate services as a 'sovereign choice', to introduce greater competition into their market (*Engineering News*, 2009). Namibia confronted a similar quandary for its exports of beef, fish and table grapes to the EU. Given the security of its own TDCA, South Africa appears to have shown insufficient regard for its smaller neighbours' peculiar challenges. By remaining outside the IEPA, South Africa also stood accused by the smaller states of pursing its parochial interests at the cost of the region, particularly SACU.

A clear lesson that emerges from the SADC EPA experience is the need for common trade and industrial policies to underpin effective regional coalition-building; but this is not always possible, given divergent interests and levels of development. In the SADC EPA, South Africa has been highly defensive. As part of a new growth path for the economy, the South African government has adopted a sector-based industrial policy, which aims to use financial incentives and sequenced tariff reforms to drive industrial development and export diversification. By contrast, some of the smaller countries are keen to diversify trade and investment away from South Africa. This may require liberalising some import tariffs and opening up competition in domestic network services sectors (e.g. energy, transport, communications and finance) (Draper and Khumalo, 2009). These differences over industrial policy initially fuelled speculation that South Africa would use the EPA conflagration to push for a 'downgrading' of SACU to an FTA. The latter would give South Africa greater policy and fiscal space (due to the common external tariff regime and revenue-sharing formula), and permit closer economic cooperation with Namibia and Angola, countries that have supported South Africa in the EPA negotiations.

The second explanation that confounded a more coordinated SADC approach involves the negotiating process itself. By all accounts, the latter was highly problematic and not adequately inclusive of national policy frameworks. In the view of one trade expert close to the negotiations: 'The end result was a chaotic, complex and challenging process, often dominated by intra-regional negotiations, tensions and mistrust as the deadline drew closer' (Kalenga, 2008). There were other complicating factors too: junior negotiators led the process, without clear mandates; national and regional priorities for the EPA were not clearly defined and divergent; and the inclusion of South Africa caused significant challenges in efforts to deal with divergent policy positions.

A related view holds that the SADC EPA's negotiating strategy and structure were largely inappropriate. Although Botswana formally coordinated the group from the outset of the process, there was no coherent or transparent regional negotiating strategy. The late inclusion of South Africa in the EPA, from an observer to participant, resulted in fundamental trade policy shifts in favour of the dominant party and the beginning of intra-group tensions and mistrust. More significantly, SADC allegedly wasted scarce negotiating time and capital on attempts to develop its own negotiating text, instead of engaging and building on the EC's text (Kalenga, 2008).

All of the above suggests that SADC's trade diplomats were not hapless victims of the EC's mercantilist machinations. Instead, their own disarray was partly responsible for the region's undoing. In other words, judicious agency still matters, particularly for small states.

The final explanation is institutional, specifically the limited involvement of the SACU and SADC Secretariats in the EPA negotiations. Within the SADC Secretariat, the roles and mandate of the EPA Unit were poorly defined and initially not well understood. As a result, the EPA Unit lacked effective capacity and authority to influence and shape the negotiations. There was also little or no collaboration between the major Southern African EPA groups (i.e. SADC and ESA) in developing negotiation strategies and positions or undertaking actual negotiations. Given that small states often lack effective negotiating capacity, regional secretariats must in future play more active roles, *inter alia* by providing the necessary support and appropriate forum for countries to exchange views, share information, generate technical analyses and policy input, define and when relevant coordinate positions, and identify best practices.

Acknowledgements

The article is based on MERCURY: Multilateralism and the EU in the Contemporary Global Order (EU FP VII, Grant Agreement Number 225267). Comments from Joseph Senona of the Department of Trade and Industry in South Africa are acknowledged. The views expressed are personal and do not reflect the position of the South African government.

Note

1. Trade deflection occurs when imports enter the FTA via the member country with the lowest tariff on non-member trade. Trade deflection distorts the region's trading patterns with the rest of the world and deprives the member country that eventually consumes the import of tariff revenue.

References

African Union (AU) High-level Panel (2007) *Audit of the African Union* (Addis Ababa: AU).

Bhagwati, J. (2008) *Termites in the Trading System. How Preferential Agreements Undermine Free Trade* (Oxford: Oxford University Press).

Bilal, S. (2009) Economic Partnership Agreements: to be or not to be, in E. Jones and D. F. Marti (Eds), *Updating Economic Partnership Agreements to Today's Global Challenges. Essays on the Future of Economic Partnership Agreements*, Economic Policy Series No. 6 (Washington, DC: The German Marshall Fund of the United States).

Carim, X. (2009) The interim EPA: view from the South African Government, in E. Jones and D. F. Marti (Eds), *Updating Economic Partnership Agreements to Today's Global Challenges. Essays on the Future of Economic Partnership Agreements*, Economic Policy Series No. 6 (Washington, DC: The German Marshall Fund of the United States).

Cooper, A. F. and Shaw, T. (Eds) (2009) *The Diplomacies of Small States. Between Vulnerability and Resilience* (Houndmills: Palgrave Macmillan).

Corporate Europe Observatory (2009) Pulling the strings of African business, 23 March, http://archive.corporateeurope.org/docs/pulling-the-strings-of-african-business.pdf, accessed 10 March 2010.

Draper, P. and Khumalo, N. (2009) The future of the Southern African Customs Union, *Trade Negotiations Insights*, 8(6), http://ictsd.org/i/news/tni/52394/, accessed 22 November 2009.

Draper, P., Halleson, D. and Alves, P. (2007) *SACU, Regional Integration and the Overlap Issue in Southern Africa: From Spaghetti to Cannelloni?* Trade Policy Report No. 15 (Johannesburg: South African Institute of International Affairs).

Elsig, M. (2009) The EU in the Doha negotiations: a conflicted leader?, in A. Narlikar and B. Vickers (Eds), *Leadership and Change in the Multilateral Trading System* (Dordrecht: Republic of Letters/Martinus Nijhoff).

Engineering News Online (2009) Economic pact with EU causes discord among Sadc members. 31 July, http://www.engineeringnews.co.za/article/i-epa-not-enhancing-regional-integration-2009-07-31, accessed 20 November 2010.

European Commission (2006) *Global Europe: Competing in the World*, Commission Staff Working Paper COM(2006)567 final.

European Commission (2010) Trade, growth and world affairs: trade policy as a core component of the EU's 2020 strategy. Communication from the Commission to the European Parliament, the Council, the European Economic and Social Committee and the Committee of the Regions, SEC(2010)1268-9.

European Services Forum (2007) ESF position paper on EU Free Trade Agreements, http://www.esf.be/pdfs/documents/position_papers/ESF%20Position%20Paper%20on%20EU%20Free%20Trade%20Agreements%20-%20final.pdf, accessed 19 January 2010.

Grobbelaar, N. and Besada, H. (Eds) (2007) *Unlocking Africa's Potential: The Role of Corporate South Africa in Strengthening Africa's Private Sector* (Johannesburg: South African Institute of International Affairs).

Jones, E. and Marti, D. F. (Eds) (2009) *Updating Economic Partnership Agreements to Today's Global Challenges. Essays on the Future of Economic Partnership Agreements*, Economic Policy Series No. 6 (Washington, DC: The German Marshall Fund of the United States).

Kalenga, P. (2008) Presentation at the TRALAC EPA Review Workshop, Cape Town, 23 January.

Lee, D. and Smith, N. J. (2008) The political economy of small African states in the WTO, *Round Table*, 97(395), pp. 259–271.

Mohammed, D. A. (2009) The CARIFORUM–EU Economic Partnership Agreement: impediment or development opportunity for CARICOM SIDS? in A. F. Cooper and T. Shaw (Eds), *The Diplomacies of Small States. Between Vulnerability and Resilience* (Basingstoke: Palgrave Macmillan).

Narlikar, A. (2003) *International Trade and Developing Countries. Bargaining Coalitions in the GATT & WTO* (London: Routledge).

Narlikar, A. (2010) *New Powers: How to Become One and How to Manage Them* (London: Hurst; New York: Columbia University Press).

Ochieng, C. (2007) The EU–ACP Economic Partnership Agreements and the 'development question': constraints and opportunities posed by Article XXIV and special and differential treatment provisions of the WTO, *Journal of International Economic Law*, 10(2), pp. 363–395.

Odell, J. S. (2000) *Negotiating the World Economy* (Ithaca, NY and London: Cornell University Press).

Oxfam (2006) Unequal partners: how EU–ACP Economic Partnership Agreements (EPAs) could harm the development prospects of many of the world's poorest countries. Oxfam Briefing Note, London, September.

Oxfam (2007) Signing away the future: how trade and investment agreements between rich and poor countries undermine development. Oxfam Briefing Paper No. 101.

Pant, M. (2009) The costs and benefits to South Africa of joining the SADC EPA. Unpublished study commissioned by the Department for International Development, London, December.

Shilimela, R. (2008) *Overlapping Memberships of Regional Economic Arrangements and EPA Configurations in Southern Africa*, FOPRISA Report No. 5 (Gaborone: Botswana Institute for Development Policy Analysis).

UNCTAD (2007) *Trade and Development Report 2007. Regional Cooperation for Development* (New York and Geneva: UNCTAD).

UNCTAD (2010) *World Investment Report 2010* (New York and Geneva: UNCTAD).

Zartman, I. W. and Rubin, J. Z. (2002) The study of power and the practice of negotiation, in I. W. Zartman and J. Z. Rubin (Eds), *Power and Negotiation* (Ann Arbor, MI: University of Michigan Press).

The Canaries in the Coalmine: Small States as Climate Change Champions

RICHARD BENWELL

Fitzwilliam College, University of Cambridge, UK

ABSTRACT *Climate change presents a useful case in the study of small states because their interests can be differentiated from larger states. Small states are expected to respond to international politics, not to lead. The development of the climate regime has seen small states engage in a 'grand strategy' to achieve climate change mitigation. The apparent powerlessness of small states and the nature of the public good problem are central to understanding small states' negotiating power in the climate regime. They have capitalised on their victim status and the common interests of all states to act as regime leaders; but they have not achieved all of their objectives in terms of access to finance and technology. This suggests that while small states share the difficulties of other developing states in pursuing value-claiming goals, they may have a comparative advantage as norm-entrepreneurs.*

Introduction

Small states are often portrayed as reactive players in international politics, forced by their relative weakness and vulnerability to external events to adjust to the changing systems around them. The best they could do, according to orthodox accounts, would be to gang up or opt out of international affairs. A recent revision of this literature has emphasised the agency of small states. Focusing on the opportunities of smallness, such as niche economic and financial activities, several revisionists have argued that small states can slip through the net of normal international relations (Cooper and Shaw, 2009; Narlikar, this issue).

One enduring characterisation of small states, however, is that they cannot successfully undertake grand political projects. Kennedy (1991, p. 186) argues that 'no doubt it is theoretically possible for a small nation to develop a grand strategy, but the latter term is generally understood to imply the endeavours of a power with extensive (i.e. not just local) interests and obligations, to reconcile its means and its end'. Size, says Jeanne Hey (2003, p. 85), 'has an absolute effect on foreign policy scope'. So, while small states can keep above the water, it is still assumed that they cannot turn the tide of international events.

In most cases, this fits with small states' tactics. In the case of climate change, however, disengagement with the systems level is not an option. The costs of independence are prohibitive (small states cannot afford adequate flood defence and unilateral mitigation or maintenance of the status quo is impossible) and failure to act poses a genuine existential threat, especially to small, low-lying states. The exigencies of climate change therefore present a new challenge for small states. For the first time, small states are at the vanguard of a political coalition that is not simply reacting to events, balancing powers or petitioning for aid; instead, small states have adopted a political position in which they are demanding that all states make substantial changes to the way they conduct their affairs. This may not be the grand strategy that Kennedy was referring to, but rallying international support to effect a fundamental change in the way that states run their economies is certainly a grand undertaking in modern political economy.

This paper suggests that climate change presents a useful case study for isolating smallness as a variable in foreign policy. As larger states are better able to adapt to the threat of climate change, they have been able to pursue more 'normal' tactics in climate negotiations, accepting delays on agreement in order to maximise their gains. Small states, by contrast, are compelled by their physical condition to pursue an 'emergency' agenda in the climate talks. Further, it is argued that the climate change dilemma creates a situation where smallness is actually a political asset in terms of allowing small states to lead in international affairs. They have achieved this through tactics of de-politicisation and appeals to science and common interests. To appreciate this point we must widen the lens we use to judge the outcomes of negotiating tactics: although small states have not achieved their value-claiming objectives of financing and technology transfer, their achievements as leading developers in a ground-breaking multilateral regime suggest that they may enjoy a comparative advantage as norm-entrepreneurs. Papua New Guinea's intervention at the Fourteenth Conference of the Parties (COP-14) in Bali, which preceded a last minute change in the US negotiating position, and the 'emergency negotiation' style adopted by Tuvalu at COP-15 in Copenhagen, which led to the suspension of the conference, was a sign of the independence and power that small states can wield in the climate regime, especially in the packed theatre of the plenary sessions.

Isolating the Size Variable

Many studies have investigated whether *smallness* has explanatory/predictive power. One of the difficulties is isolating smallness as an explanatory variable from other variables, such as *poorness*. The climate change case offers a route past this problem because the urgency associated with mitigation is amplified by smallness to the level of an existential threat. While underdevelopment and poverty also amplify the impacts of climate change, increasing the likely costs of delayed mitigation, they do not necessarily lead to an existential security challenge. Likely impacts are in the range of several to tens of per cent of GDP in developing countries in general (Alliance of Small Island Developing States (AOSIS), 2008b, p. 1), but for small states the effects of climate change can equal 200% of GDP, even from a single hydrometeorological event (AOSIS, 2008a, para. 10; AOSIS, 2008b, p. 2). This creates a dividing line between the tactics that *small, poor* states are pursuing in climate negotiations and those pursued by *just poor (larger)* states and *large, rich* states.

For large, poor states, an entrenched bargaining position is possible because the options for adaptation to climate change are wider and the likely effects less catastrophic. Larger states are therefore able to balance the need to mitigate and adapt to climate change with other concerns, such as development and competitiveness. While the exigency of the situation has prompted small island developing states (SIDS) to call for nationally appropriate mitigation actions (NAMAs) *including* for developing countries, prominent larger developing states such as India and China continue to hold the line set in the run up to the Framework Convention that developing states should not be obliged to take on quantified mitigation targets (Depledge, 2000, para. 476; *Earth Negotiations Bulletin*, 2009a, p. 4). Large, poor states are willing to accept the costs of delayed action on climate change in order to secure a deal that wins relative benefits and room for development.

By contrast, small states argue that action on climate change must be concerted and urgent and are less willing to compromise than their larger counterparts. For small islands in particular, development and competitiveness cannot be considered separately from the issue of surviving climate change. This has produced both internal focus and mutual cooperativeness between small islands. The US Department of State (2009) notes, for example, that the promotion of concern about anthropogenic sea-level rise is Tuvalu's 'major international priority'. Early in the negotiation process, therefore, small island states formed a separate group from the G-77/China coalition of developing states.

The G-77 and China brings together a coalition based on development, including small, poor countries alongside large, rapidly developing countries such as India (Yamin and Depledge, 2004, p. 35). AOSIS, on the other hand, is a grouping of 39 states (and four observers) based around physical attributes, not affluence. Small states are galvanised to act together in response to a systemic threat, conforming to the classic characterisation of small states as particularly susceptible to systems-level change (Hey, 2003, pp. 186–187). Whereas larger states have the capacity to delay their response to that threat, small states are rendered unable to do so by their physical attributes.

Small island states have maintained coalitions before: the South Pacific Commission has functioned since 1946, for instance. The Commission's agenda, though, has been much more regional, concentrating on localised issues such as fisheries (Chasek, 2005, p. 14). AOSIS is notable for combining the Caribbean, Pacific, and the African, Indian Ocean and Mediterranean small island/low-lying coastal states and for its global sphere of activity. Some small states are members of AOSIS even though they are not UN members. In these circumstances, at least, smallness itself can be a unifying factor even where in other circumstances the states seem to have little in common (AOSIS, 2000).

Small states have also exerted their shared purpose in other forums such as the Commonwealth, where they have used their numerical strength in the Heads of Government meetings to steer the agenda. In 2007, paragraph 3 of the Commonwealth Climate Change Action Plan recognised climate change as 'a direct threat to the very survival of some Commonwealth countries, notably small island states'. In the Climate Change Consensus, Commonwealth Heads of Government (CHOGM) 2009 reiterated the existential threat (para. 1) and pointed out that small, low-lying states and less-developed countries (LDCs) 'face the greatest challenges, yet have contributed least to the problem of climate change' (para. 3). Both documents emphasise 'collective, comprehensive and global action' (CHOGM, 2007, para. 4, 2009, para. 8), underlining the focus among small states and the Commonwealth on promoting international consensus to solve the public good problem of mitigation.

This unity based on smallness and vulnerability is manifest in the relatively *physical* targets proposed by AOSIS throughout the negotiations. The Alliance promoted the 2°C limitation of increase above pre-industrial temperatures and—now that 2°C has been taken up by the EU and others—has now revised the figure down to 1.5°C, even beyond Intergovernmental Panel on Climate Change (IPCC) recommendations. Similarly, AOSIS argued for a cap of atmospheric greenhouse gases (GHGs) concentrations at 450 ppm before revising the figure to 350 ppm. They have also suggested thresholds for sea-level rise and the avoidance of adverse effects on SIDS as a benchmark for climate change mitigation success (FCCC/AGBM/1996/MISC.2; AOSIS, 2000, 2009). These targets emphasise that, for small states, what can sometimes seem an abstract and remote threat is in fact a proximate physical threat to their security and that physicality underlines the cooperation between these states.

A third group comprises rich, small states. Rich, small states are able to afford greater access to adaptation technologies than small, poor states and the urgency of the threat is reduced. Even rich, small states such as the Netherlands are threatened by climate change, but their ability to invest in adaptation measures tempers the danger; lower levels of development increase the prominence of environmental security as a foreign policy goal (Hey, 2003, pp. 193–194; Cooper and Shaw, 2009, p. 31).

The negotiating position of these groups has been significantly different during the Intergovernmental Negotiating Committee, the Berlin Mandate and more recently the Bali Action Plan negotiations, differentiated by whether they can afford to play negotiating games that may delay action on climate mitigation, as they would in other talks such as trade negotiations. For larger developing states, responsibility and historical emissions are a sticking factor in talks; for small island states, this dynamic is overridden by the survival instinct and creates a relatively unified interest group.

Achievements

Small states have achieved many successes both in political processes and in policy outcomes, including *legal* achievements (both value-claiming and process achievements) and *implementation* achievements, in the actual mitigation activities undertaken by other states.

Several authors have concluded that small states have been unsuccessful in their attempts to influence the climate regime. They argue that small states' lobbying has successfully raised awareness, but not enough to fulfil underlying goals (Chasek, 2005, p. 135; Baldacchino in Cooper and Shaw, 2009, p. 35). These judgements, however, rely on value-claiming assessments of success, based on achievements in other issue areas such as aid budgets. Contrary to common measures of negotiating success, however, the success of small states in the climate regime should be seen in terms of overall mitigation action. It is the objective of a 'grand political movement' to make other states change their own behaviour and policies.

Value-claiming Achievements

The climate regime includes exceptions, exemptions and extra assistance for small states at each political level. It is unlikely that the special needs of small island and low-lying states for adaptation and capacity-building would have been recognised without the concerted lobbying of AOSIS (Larson, 2003, pp. 144–145); but these value-claiming achievements, although substantial, are far short of the resources necessary to sustain small states in the long term.

FCCC Article 4(8)(a) obliges Parties to consider funding, insurance and the transfer of technology to help to meet 'the specific needs and concerns of developing country Parties ... especially ... small island countries'. This was reiterated in the Marrakesh Accords in 2001, and Decisions at COP-7 repeated the formula of requiring financial, technical assistance for the most vulnerable. Decision 6/CP.7, paragraph 1, went on to enact this principle in the guidance to the Framework Convention on Climate Change (FCCC) financial mechanism. Similarly, Decision 2/CP.7 highlights the need for capacity-building for LDCs and SIDs to be funded by Annex II states. The Global Environmental Facility (GEF) now runs adaptation facilities that contribute to funding small states' adaptation activities. Alongside GEF funding, vulnerable countries are afforded a revenue stream from a share of the proceeds from the Clean Development Mechanism under Article 12 of the Kyoto Protocol, channelled through the Adaptation Fund.

In total, however, the level of finance that has been pledged and delivered for small states' adaptation has been far below their requirements, even 15 years after the Global Conference on the Sustainable Development of SIDS. By 2008, the GEF Strategic Priority on Adaptation, the Least Developed Country Fund and the Special Climate change Fund had raised only US$280m put together (AOSIS, 2008b, p. 30). The 'goal' of US$100bn in finance set out in the Copenhagen Accord is only half the amount small islands proposed developed states should commit to in their Copenhagen Protocol proposal.

Process Achievements

> The principle of representation proportionate to risk struck a sympathetic chord at the Second World Climate Conference. (AOSIS, 2000)

The story of small states' institutional involvement in the climate regime begins from the usual position of lack of capacity: it is difficult for nations with fewer international diplomats and experts to field a team that can participate fully and effectively in complex multilateral negotiations (Hey, 2003, p. 190; Roberts and Parks, 2007, p. 15).

The first major achievement of small islands was to ensure their recognition as a distinct category in the UN. The term small island developing states was coined in 1992 at the Rio Conference and their special environmental and developmental needs are acknowledged in Chapter 17 (G) of Agenda 21.

The most important institutional *process* achievement by small states has been to secure a special seat on the COP Bureau, alongside the five UN regional groupings. This decision, under COP Rule 22, was an innovation under the climate regime in recognition of the right to representation of this high-risk group (Yamin and Depledge, 2004, p. 412). Furthermore, this decision has created a precedent for SIDS' representation in other committees and groups, such as the two Kyoto compliance Branches, the CDM Executive Board, the Adaptation Fund and even on the Supervisory Committee for Joint Implementation, a mechanism that involves only Annex I states.

Small states have also often provided chairmen and vice-chairmen (Yamin and Depledge, 2004, p. 420). AOSIS was entrusted with leadership of the Intergovern-mental Negotiating Committee (INC) established to draft the Convention. Michael Zammit Cutajar from Malta chairs the crucial Ad Hoc Working Group on Long-term Cooperative Action (AWG-LCA), which was mandated by the Bali Action Plan to focus on four key elements of long-term cooperation identified during the Convention Dialogue.

Wider Achievements

Value-claiming achievements are important to small states to support adaptation measures to cope with the effects of climate change and to help them to achieve their Millennium Development Goals. For the smallest states, however, adaptation can provide only a temporary reprieve. As Tuvalu's negotiator Ian Fry remarked at COP-15, 'All of the billions and trillions in the world won't do a darn thing if your country is drowning or, worse yet, no longer exists' (Solutions, 2009). The ability to influence the progress of the wider climate change regime should be counted as a mark of the power and influence of small states.

The Framework Convention provides the backbone of the climate regime and, by winning the inclusion of important normative principles, small states and other developing countries set a foundation for all subsequent discussions. The principles laid out in the Preamble and Article 3 are equity; 'common but differentiated responsibilities' (Article 3(1)); the precautionary principle (Article 3(2)); and sustainable development (Article 3(4)).

During the FCCC negotiations, one of the less successful objectives pursued by AOSIS was the promotion of an insurance fund for offsetting climate impacts. In the end, the reference to 'insurance' in Article 4.8 was the only result of AOSIS's 1991 suggestion of an international insurance pool (Yamin and Depledge, 2004, p. 227). In 2005, however, this ambition was achieved outside the FCCC through the World Bank Catastrophe Risk Insurance Facility (CRIF). The world's first-ever multi-country catastrophe insurance pool, US$47m, was pledged in the first donation round (AOSIS, 2008b, p. 32).

Most importantly, small states have consistently been innovators in the negotiation process. They have worked in partnership with environmental non-governmental organisations (NGOs) and lobby groups to update consistently and pioneer target-setting, normative issues and legal innovation. In 1988, Malta introduced climate change to the UN General Assembly (UNGA) in resolution 43/53, which recognised climate change as a common concern of mankind. Following this, in June 2009, at the instigation of the SIDS, climate change was recognised by UNGA as a threat to global security.

On behalf of AOSIS, Trinidad and Tobago was the first Party to submit a draft protocol under the Berlin Mandate (FCCC/AGBM/1996/MISC.2), giving small island states a first-mover advantage in setting the terms of the protocol to the FCCC. Once the Protocol had been finalised, the Maldives was the first country to sign it. Then, in 2003, it was AOSIS that insisted that climate talks should continue as scheduled, despite suggestions by the EU that they should be delayed (*The Guardian*, 2003).

In 2009, small states continued to be active participants in the COP/MOP. Typically, as part of their 'emergency' negotiation style, AOSIS themselves were slightly ahead of the 'next most radical' negotiating position (often held by the EU), for example updating their temperature and GHG concentration targets to 1.5°C and 350 ppm. They have once again created momentum and unlocked turgid talks by releasing their own legal texts, such as the *Proposal for the survival of the Kyoto Protocol* and the *Copenhagen Protocol*, which build on the Bali Action Plan to integrate US and developing country commitments into the Kyoto Protocol legal framework. This double agenda proposed the extension of the Kyoto Protocol to 2017 and the creation of a Copenhagen Protocol, bringing the US on board and committing developing states to Nationally Appropriate Mitigation Actions.

Key members of AOSIS have always been frontrunners, supporting the Alliance position but separately advocating further action; this continued in Copenhagen with the Tuvaluan draft protocol (FCCC/CP/2009/4). The Tuvaluan proposal (which came first) would also have amended the Kyoto Protocol to form an Annex BI, which would have allowed non-Annex I developing states to take on commitments under KP Annex B, honing in on a key legal obstacle in the Protocol (Depledge and Yamin, 2009, p. 445).

The most dramatic and overtly effective negotiating tactics adopted by small states have been their theatrical interventions in the COP plenary sessions. At COP-14 in Bali, on behalf of Papua New Guinea, Kevin Conrad entreated 'I would ask the United States, we ask for your leadership. But if for some reason you're not willing to lead, leave it to the rest of us. Please get out of the way'. Minutes later, United States negotiator Paula J. Dobriansky ended her objection to the Bali Roadmap and stated 'We will go forward and join consensus today' (*New York Times*, 2008). A year later in Copenhagen, Tuvalu pursued similar grandstand tactics. Supported by other small island states, including Grenada, Trinidad and Tobago and several African states, the small states parted company with their traditional negotiating allies China, Saudi Arabia and India. Their proposal, which became known as the 'Tuvalu Copenhagen Protocol', called for a legally binding agreement that would hold developing countries accountable for their emissions alongside the Annex I states (*The Guardian*, 2009). The proposal was rejected, but Tuvalu (with a population of 12,000) caused the suspension of the Conference of the Parties and disrupted attempts by other Parties to present the non-legally binding Copenhagen Accord as a success. With plenary sessions broadcast around the world, the small island states were supported by powerful NGO rallies. Their emotional rhetoric and capture of formal rules of negotiation procedure combined in these circumstances to make them powerful independent players in the climate change regime.

What's New?

Together, these achievements mark an important contribution by small states to the advancement of the 'grand strategy' of creating an effective climate mitigation regime.

This ability to influence the course of high politics is at odds with the notion that small states lack power and influence in international affairs. Jeanne Hey (2003, p. 5), for example, has summarised the literature on small states, arguing that small states often:

- exhibit a low level of participation in world affairs;
- address a narrow scope of foreign policy issues;
- limit their behaviour to their immediate geographic arena;
- employ diplomatic and economic foreign policy instruments, as opposed to military instruments;
- emphasise internationalist principles, international law, and other 'morally minded' ideals;
- secure multinational agreements and join multinational institutions whenever possible;
- choose neutral positions;
- rely on superpowers for protection, partnerships and resources;
- aim to cooperate and to avoid conflict with others;
- spend a disproportionate amount of foreign policy resources on ensuring physical and political security and survival.

In the case of climate change mitigation, however, small states behave very differently in three key measures in this list. First, small states do not 'exhibit a low level of participation'; on the contrary, they have been vocal advocates of swift mitigation efforts. Second, they do not 'limit their behaviour to immediate geographic arena', but campaign at the global level. Third, small states do not 'choose neutral positions', rather they have adopted the most progressive political position expressed at international climate change talks.

As Hey notes, not all of these small state characteristics are present all of the time, but in the case of climate change these three differences represent a noteworthy departure from the normal behaviour of small states.

The key to explaining this departure is threefold:

- the nature of the climate change threat as a global public good problem means that the vector of international politics is towards cooperation;
- small states' rhetoric and cohesion are empowered by the threat of climate change as an existential environmental security issue;
- small states' negligible contribution to the problem combines with the likelihood that they will be worst affected to give them victim status in international negotiations, allowing them to act 'above politics'.

These factors create a situation where small states emerge as natural leaders in climate politics.

Small States and Effective Leadership

The small States of the world ... are more than capable of holding their own ... their contributions are the very glue of progressive international cooperation for the common good. (Kofi Annan, 1998, UN Secretary-General)

As Kofi Annan recognised in 1998, contrary to the view set out by Hey, global leadership is exactly what small states have to offer in certain issue areas, principally where there is the necessity of 'cooperation for the common good'.

Leadership in international affairs is not a simple translation of power to outcomes. Considerations such as moral standing are important, as demonstrated by the disillusionment with Western agendas in fields such as trade, environment and security (Hill, 2009, pp. 20–21). The realisation that the human race has the potential for self-destruction has changed the philosophical context in which international politics takes place (Hill, 2009, p. 13). In these circumstances, although the size and wealth of 'big' states remain key to the bargaining process, the interplay of power politics with economic competitiveness can limit the sort of soft power necessary for the success of collaborative leadership (Nye, 2008, pp. 43–50). In a public good dilemma, no one state can manage unilaterally, while the perception that a party is positioning for relative advantage can undermine its potential to lead. It is the vulnerability of small states that sets them apart from political wrangling and gives them a natural leadership role.

Neo-realist understanding puts the power in resources. The problem is compounded with a wider understanding of power. Structural power is based on a combination of institutional and economic advantages such as GDP and trade volumes. In the climate talks, an effective bargaining chip is a promissory note to achieve a certain emissions reduction, and those with the biggest stack of chips are those who pollute the most. This dynamic took hold in the talks under the Bali Action Plan. States attempted to influence the resolution of climate commitments by laying offered reductions on the table. In some cases, the extent of reductions was conditional on the reductions offered by other Parties (*Earth Negotiations Bulletin*, 2009b, p. 3). Small states contribute little to the climate change threat and so can bring little to the bargaining table in terms of promised reductions.

In the climate regime, small states' main resource is the power of exhortation based on: warnings concerning an impending tragedy of the commons; historical responsibility; scientific principles; sovereign states' legal and moral right to existence; and the right to sustainable development. The theme that unites these discursive strands is that as the principal victim of a common resource problem not of their own making, small states' power lies in their powerlessness. Baillie (1998) argues that in the EU, Luxembourg has capitalised on the perception that it has no hegemonic aspirations to maximise its influence. In this way, small states cannot be perceived as jockeying for relative gains in climate negotiations. Whereas Luxembourg's aspirations are regional, however, small states cannot confine themselves to local issues. Instead, they have positioned themselves to play a role as a cautionary tale in international politics.

The Canaries in the Coalmine

Imagine the world knew global warming was about to destroy 43 nations—but not which 43. (Enele Soponga, UN Ambassador of Tuvalu, in *Sydney Morning Herald*, 2006)

The unifying theme in small states' advocacy of climate change mitigation is based on the nature of the climate change as a global public good problem. Small states present themselves not as value-claiming parties vying for relative gain, but as 'the canaries in the coalmine' in a tragedy of the commons that is unfolding for all states. SIDS have referred to themselves as 'front-line states', alluding to their political leadership and to their position as first victims of climate change. In the context of consensus about the necessity for action, small states can act as a catalyst, especially through coalition work in international organisations (Hey, 2003, p. 188). There is an understanding among states that cooperation will have to be achieved *at some point* to avert a crisis. The direction of political movement is thus towards cooperation; delays ensue as a result of competing for advantages *en route* to a settlement.

In order to convey this message of the commonality of the threat, small states have sustained a normative pressure throughout negotiations, harnessing scientific inputs, international law and interest-based bargaining strategies to catalyse the advancement of the regime. In the formative years of the regime, small islands contributed to the principles and rules that guide the negotiating process, earning special recognition on the basis of the existential threat that climate change poses and helping to tighten the mitigation objectives of the Kyoto Protocol. In more recent years, their 'emergency' negotiating style has taken advantage of the theatre of the plenary sessions and the COP procedural rules to become powerful players, independently of the EU and the G-77 negotiating bloc.

The science that informed the environmental movement was mostly the work of developed countries. Small states have, however, been among the canniest users of scientific understanding to support their case and the strongest advocates of science-based policy and advanced knowledge in specific scientific fields. In 2005, for instance, Seychelles President James Michel helped to institute the Global Islands Partnership (GLISPA) against sea-level rise and the Sea-level Rise Foundation, a global initiative to establish a platform of excellence on sea-level rise. Utilising science in this way is part of the 'depoliticising' strategy pursued by small states to emphasise that mitigation is in the rational interests of all states as a 'common concern of mankind' (UNFCCC preamble).

In order to underscore their message of a common threat, small states have emphasised their status as victims (Gayoom, 1987; AOSIS, 1999, 2000, 2009; Koonjul, 2004; *Bangkok Post*, 2009). AOSIS has emphasised historical responsibility as a guiding principle in climate talks, drawing on norms of equity and fairness and the polluter pays principle. Invoking the precautionary principle, they argue that this vulnerability means that they will only be the first states to suffer if action is not taken. Charles Fleming, the Ambassador of St Lucia, warned that 'if we wait for the proof, the proof will kill us' (cited in AOSIS, 2000). In 2009, President Mohamed Nasheed decided that the Maldives would not send a delegation to COP-15 in order to save money, which could be spent on a replacement homeland (*Bangkok Post*, 2009), and instead would hold an underwater cabinet meeting in order to highlight the threat of sea-level rise (UN COP-15 website 2009). Unlike the broader G-77/China coalition, AOSIS have not compromised this moral authority by insisting that lack of historical responsibility means that they should not take on any targets, instead advocating voluntary objectives for developing nations.

The Future of the Climate Regime

In 2011, the future of the Kyoto Protocol is in doubt. The Protocol's first commitment period ends in 2012 and, unless new periods are set, the Protocol will expire. As a result, investment in the Clean Development Mechanism (CDM) is increasingly risky, as businesses begin to doubt whether Certified Emissions Reductions (CER) revenue streams will materialise. At the COP-15, the US delegation argued for an alternative to Kyoto and was rewarded with the Copenhagen Accord, a document drafted by a small club—the BASIC countries, Brazil, South Africa, India and China, and the US—in talks that excluded other groups such as the EU and AOSIS. The Accord aims for a 2°C limit, not the 1.5°C and 350 ppm targets that AOSIS wanted. Moreover, as a 'politically binding' deal based on unilateral pledges, it is very different from the legally binding commitments based on the Kyoto Protocol legal structure that AOSIS advocated. There is little hope for a binding deal in Cancún at COP-16.

Under these circumstances, the unity maintained in the formative period of the climate regime has begun to erode. During the COP-15 negotiations, although Tuvalu was able to have a powerful influence on the course of the Conference, a new dynamic emerged with members explicitly opposing the Alliance position, for example when Papua New Guinea supported Brazilian compromise proposals instead of the legally binding Copenhagen Protocol proposed by AOSIS (Climate Pacific, 2009). It remains to be seen whether this was an isolated event or a sign of weakness in the Alliance.

The small island states remain in grave danger from climate change. One of the prevailing political dynamics of the climate regime remains the zero-sum value-claiming competition between a coterie of rich countries and a group of major developing countries, with members of OPEC (Organisation of the Petroleum Exporting Countries) resisting action even more strongly. Much of the headway that AOSIS and other climate leaders have already made stands to be lost should the Kyoto Protocol be abandoned, as the US would prefer (*Earth Negotiations Bulletin*, 2009a, p. 1).

The success of the regime will be partially contingent on small states' ability to continue to pursue their emergency agenda 'above politics' as norm-entrepreneurs. Although small states have not fulfilled all of their value-claiming aspirations in the climate arena, such as financing and technology transfer, they have been able to promote the development of the mitigation regime itself: a 'grand strategy' in political economy. Small states' ability to influence the regime is a function of their smallness. They are gifted this paradoxical 'power of the powerless' through the commonality of the problem and their victim status, which help to raise them above the value-claiming behaviour of politics-as-usual. This suggests that small states may enjoy a comparative advantage as norm-entrepreneurs in international environmental politics, which may be partially transferable to other issue areas.

References

Annan, K. (1998) Secretary-General lauds role of small countries in work of United Nations, noting crucial contributions, SG/SM/6639, 15 July.

AOSIS (1999) Communiqué: Third Summit of the Heads of State and Government of the Alliance of Small Island States (AOSIS), 25 September.

AOSIS (2000) Climate change and small island states, Ministerial Conference on Environment and Development in Asia and the Pacific, Kitakyushu, Japan, 31 August–5 September.

AOSIS (2008a) Addressing climate change: the United Nations and the world at work, UN General Assembly Thematic Debate, 11–12 February.

AOSIS (2008b) Global climate change and small island developing states: financing adaptation, Draft Green Paper, 5 February.

AOSIS (2009) Declaration on climate change, High Level Summit on Climate Change, New York, 21 September.

Baillie, S. (1998) A theory of small state influence in the European Union, *Journal of International Relations and Development*, 1(3/4), pp. 195–219.

Bangkok Post (2009) Maldives 'too broke' to attend climate summit, 7 September.

Brown, P. (2003) Drowning islands halt effort to postpone climate change talks, *The Guardian*, 13 December.

Button, J. (2006) Tiny Tuvalu packs a powerful punch, *Sydney Morning Herald*, 20 November.

Chasek, P. S. (2005) Margins of power, International Studies Association 46th Annual Convention, Honolulu, 1–5 March.

Climate Pacific (2009) Papua New Guinea breaks ranks with AOSIS and supports Brazil, ClimatePacific@COP-15, 12 December.

Commonwealth Heads of Government Meeting (2007) *Lake Victoria Commonwealth Climate Change Action Plan*, Kampala, Uganda, 24 November.

Commonwealth Heads of Government Meeting (2009) *Port of Spain Climate Change Consensus: The Commonwealth Climate Change Declaration*, Trinidad and Tobago, 28 November.

Cooper, A. F. and Shaw, T. M. (2009) *The Diplomacies of Small States: Between Vulnerability and Resilience*, International Political Economy Series (Basingstoke: Palgrave Macmillan).

Depledge, J. (2000) *Tracing the Origins of the Kyoto Protocol: An Article-by-Article Textual History*, Technical Paper, FCCC/TP/2000/2, 25 November.

Depledge, J. and Yamin, F. (2009) The global climate-change regime: a defence, in D. Helm and C. Hepburn (Eds), *The Economics and Politics of Climate Change* (Oxford: Oxford University Press), pp. 433–453.

Earth Negotiations Bulletin (2009a) AWG-LCA 7 and AWG-KP 9 highlights: Monday, 28 September 2009, in T. Akanle *et al.*, International Institute for Sustainable Development, 12(429), 29 September.

Earth Negotiations Bulletin (2009b) Summary of the Bonn climate change talks: 10–14 August 2009, in A. Appleton *et al.*, International Institute for Sustainable Development, 12(427), 17 August.

Gayoom, M. A. (1987) Address by His Excellency Mr. Maumoon Abdul Gayoom, President of the Republic of Maldives, before the Forty Second Session of the United Nations General Assembly on the Special Debate on Environment and Development, 19 October.

Hey, J. A. K. (Ed.) (2003) *Small States in World Politics: Explaining Foreign Policy Behavior* (London: Lynne Rienner).

Hill, C. (2009) Changing the world? Paper presented at the Department of Politics and International Studies (POLIS), University of Cambridge, 5 May.

Kennedy, P. (1991) Grand strategy in war and peace: toward a broader definition, in P. Kennedy (Ed.), *Grand Strategies in War and Peace* (New Haven, CT: Yale University Press), pp. 1–7.

Koonjul, J. (2004) Remarks by Ambassador Jagdish Koonjul Permanent Representative of Mauritius and Chairman of AOSIS at the Forum of Small States, Washington, DC, 3 October.

Larson, M. J. (2003) Low-power contributions in multilateral negotiations: a framework analysis, *Negotiation Journal*, 19(2), pp. 133–149.

Nye, J. S. (2008) *The Powers to Lead* (Oxford: Oxford University Press).

Revkin, A. C. (2008) Issuing a bold challenge to the U.S. over climate, *New York Times*, 22 January.

Roberts, J. T. and Parks, B. C. (2007) *A Climate of Injustice: Global Inequality, North–South Politics, and Climate Policy* (Cambridge, MA: MIT Press).

Solutions (2009) COP 15: accepting responsibility, *Solutions Journal Online*, https://www.the solutionsjournal.com/node/518#comment-87

UN COP-15 website (2009) http://www.denmark.dk/en/menu/Climate-Energy/COP15-Copenhagen-2009/cop15.htm

US Department of State (2009) Background note: Tuvalu, Bureau of East Asian and Pacific Affairs.

Vidal, J. (2009) Copenhagen talks break down as developing nations split over 'Tuvalu' protocol, *The Guardian*, 9 December.

Yamin, F. and Depledge, J. (2004) *The International Climate Change Regime* (Cambridge: Cambridge University Press).

Index

Page numbers in **bold** type refer to figures
Page numbers in *italic* type refer to tables